SpringerBriefs in Environmental Science

SpringerBriefs in Environmental Science present concise summaries of cutting-edge research and practical applications across a wide spectrum of environmental fields, with fast turnaround time to publication. Featuring compact volumes of 50 to 125 pages, the series covers a range of content from professional to academic. Monographs of new material are considered for the SpringerBriefs in Environmental Science series.

Typical topics might include: a timely report of state-of-the-art analytical techniques, a bridge between new research results, as published in journal articles and a contextual literature review, a snapshot of a hot or emerging topic, an in-depth case study or technical example, a presentation of core concepts that students must understand in order to make independent contributions, best practices or protocols to be followed, a series of short case studies/debates highlighting a specific angle.

SpringerBriefs in Environmental Science allow authors to present their ideas and readers to absorb them with minimal time investment. Both solicited and unsolicited manuscripts are considered for publication.

Lisa F. Clark • Jill E. Hobbs

International Regulation of Gene Editing Technologies in Crops

Current Status and Future Trends

 Springer

Lisa F. Clark
Department of Agricultural and Resource
Economics
University of Saskatchewan
Saskatoon, SK, Canada

Jill E. Hobbs
Department of Agricultural and Resource
Economics
University of Saskatchewan
Saskatoon, SK, Canada

ISSN 2191-5547   ISSN 2191-5555   (electronic)
ISBN 978-3-031-63916-6   ISBN 978-3-031-63917-3   (eBook)
https://doi.org/10.1007/978-3-031-63917-3

This Springer imprint is published by the registered company Springer Nature Switzerland AG
The registered company address is: Gewerbestrasse 11, 6330 Cham, Switzerland

If disposing of this product, please recycle the paper.

Acknowledgements

We would like to thank all the interviewees who gave generously of their time and expertise and whose insights were invaluable in informing this manuscript. We would like to thank Dr. Peter W.B. Phillips for his regulatory policy insights and Dr. Stuart Smyth for his helpful suggestions regarding interview questions. We thank Dr. Sateesh Kagale for his feedback on Chaps. 2 and 4. Any errors or omissions remain those of the authors.

Funding acknowledgement: This research was conducted as part of the 4DWheat: Diversity, Domestication, Discovery, and Delivery project. The authors acknowledge funding provided by the '4D Wheat: Diversity, Domestication, Discovery and Delivery' project, funded by Genome Canada, Agriculture and Agri-Food Canada, Western Grains Research Foundation, Saskatchewan Ministry of Agriculture, Saskatchewan Wheat Development Commission, Alberta Wheat Commission, Manitoba Crop Alliance, Ontario Research Fund, Viterra, Canadian Agricultural Partnership, and Illumina. We also acknowledge the administrative support of Genome Prairie.

Declaration of Competing Interests

The authors have no conflicts of interests to declare that are relevant to the content of this manuscript.

Ethics Approval

This research received ethics approval from the University of Saskatchewan Behavioural Research Ethics Board (BEH 3922) in March 2023. Informed consent to participate in this research was obtained from all participants, as per this ethics approval.

Contents

List of Figures

List of Tables

About the Authors

Lisa F. Clark   is a Research Associate with the Department of Agricultural and Resource Economics, University of Saskatchewan. She holds a Ph.D. in Political Science from Simon Fraser University, with specializations in public policy and international relations. Her research interests include innovative technologies in the agrifood system, food policy, food quality standards and labelling, and food security. Lisa has published research on the organic food sector, plant-based proteins, global food security, as well as geographical indications and climate change. Her forthcoming publications focus on indigenous agriculture and innovations in Arctic food systems.

Jill E. Hobbs   is a Professor in the Department of Agricultural and Resource Economics, University of Saskatchewan. She holds a Ph.D. from the University of Aberdeen and is a Fellow of the Canadian Agricultural Economics Society. Her research interests include agri-food supply chains, food policy, and consumer behaviour. She has published widely on topics ranging from consumer responses to new food technologies, the regulation of health foods, public and private standards for food safety and food quality, and supply chain resilience in the agri-food sector.

Abbreviations

TRIPS	Agreement on Trade-Related Aspects of Intellectual Property Rights
AAFC	Agriculture and Agrifood Canada
AIS	Agriculture Innovation Systems
AFD	Animal Feed Division
CFIA	Canadian Food Inspection Agency
CPB	Cartagena Protocol on Biosafety
CRISPR	Clustered Regularly Interspaced Short Palindromic Repeats
CBD	Convention on Biological Diversity
DNA	Deoxyribonucleic Acid
DSB	Double-Strand Break
EFSA	European Food Safety Authority
EU	European Union
EC	European Commission
FAO/WHO	Food and Agriculture Organization/World Health Organization
FSANZ	Food Standards Australia New Zealand
FTO	Freedom to Operate
GABA	Gamma-Aminobutyric Acid
GMO	Genetically Modified Organism
HDR	Homology-Directed Repair
INTA	Instituto Nacional de Tecnologia Agropecuaria (Agricultural Technology of Argentina)
IP	Intellectual Property
IPR	Intellectual Property Rights
IPPC	International Plant Protection Convention
LMO	Living Modified Organism
NASEM	National Academies of Sciences, Engineering, and Medicine
NBT	New Breeding Techniques
NHEJ	Non-Homologous End-Joining
NSF	National Science Foundation
ODM	Oligo-Directed Mutagenesis
OECD	Organisation for Economic Co-operation and Development

PBO	Plant Biosafety Office
PPE	Plant Prime Editing
PAM	Protospacer Adjacent Motif Sequence
PRRI	Public Research and Relations Initiative
RNA	Ribonucleic Acid
RNAi	Ribonucleic Acid interference
SPSMs	Sanitary and Phytosanitary Measures
STOA	Scientific Technology Options Assessment
SDN	Site-Directed Nucleases
TALENs	Transcription Activator-Like Nucleases
WIPO	World Intellectual Property Organization
WOAH	World Organisation for Animal Health
WTO	World Trade Organization
US	United States of America
USD	United States Dollars
ZFN	Zinc Finger Nuclease

Part I
Balancing Innovation: Application and Regulation

Chapter 1
Introduction: Why Study the Governance of Gene Edited Agrifoods?

1.1 Introduction

In 2012, a research team led by Jennifer Doudna including Emmanuelle Charpentier, Martin Jinek, Krzysztof Chyliński, Ines Fonfara and Michael Hauer at UC Davis published their findings that would revolutionize gene editing. They discovered that a protein (Cas9) within an organism's cell could be programmed with RNA to edit its genomic DNA. In essence, the CRISPR-Cas9 technique harnessed a cell's own repair mechanisms to make precise gene edits. When the findings were published, CRISPR's potential and possibilities took the scientific world by storm. Unlike earlier gene editing techniques, CRISPR could perform gene editing much faster and was much more cost effective. Doudna and Charpentier would go on to win the Nobel Prize in Chemistry in 2020 for their discovery. Some have called this one of the most important discoveries in the history of biology.

The promise of CRISPR to improve the precision of plant breeding and the potential to find genetic treatments and cures for human diseases is staggering. But the initial excitement about the future benefits of gene editing in agriculture took a very different turn in 2018. A conference held in Hong Kong brought scientists together from around the world to discuss the implications of using CRISPR-Cas9 to edit genes for human health research as well as agricultural applications. During the conference, a researcher from China made a presentation that would leave a lasting impression. In this presentation, scientist He Jiankui announced that CRISPR gene editing had been used successfully on twin girls to make them resistant to HIV/AIDS infections. At the time, there were no regulations or policies in place to prevent ethically questionable uses of CRISPR-Cas9 to alter the genome of any organism, including humans. The global controversy and condemnation by the scientific community surrounding this announcement triggered a strong regulatory response in countries around the world (Asquer & Krachkovskaya, 2021: 1117).

L. F. Clark, J. E. Hobbs, *International Regulation of Gene Editing Technologies in Crops*, SpringerBriefs in Environmental Science, https://doi.org/10.1007/978-3-031-63917-3_1

The 2018 event would have lasting effects on the trajectory of gene editing in agricultural research up until today.

The introduction of innovative technologies into the world often requires what Klerkx et al. (2010: 391) call 'selling a good story'. The story must be told by the right people who have credibility, at the right time, in the right place and to the right people who can 'capitalize on momentum and windows of opportunity'. If not, fear, uncertainty and outrage may dominate the narrative. It can easily be argued that gene editing, and particularly CRISPR-Cas9, was not the subject of a good story. It was instead, introduced to the world as an ethically dubious technology wielded by unscrupulous scientists to re-write the human genome without any social license to do so. The legacy of the focusing event in 2018 has had long standing implications for gene editing for research into its applications for human health, but also for how this innovation has been welcomed or resisted in the agricultural innovation space.

The governance of gene editing in agricultural research may seem unrelated to this event, but the resultant regulatory wheels that were set in motion have had significant implications for the regulation and use of gene editing in the global agrifood system. On one hand, proponents of gene editing techniques in agriculture argue that it should not be considered identical to genetically modified foods because it does not use transgenes (the SDN1 technique) and thus, should be regulated similarly to any other agrifood introduced to a country with no prior safety record—genetically edited or produced using conventional techniques. On the other, gene editing is viewed by some as 'just another biotechnology' that carries the same perceived risks as Genetically Modified Organisms (GMOs)[1] and should be subject to the same biosafety protocols and regulations as its genetically modified counterpart. Currently, there is no global consensus on how gene editing should be regulated in agrifood research or assessed for commercialization.

Gene editing has been used in agricultural research and product development for decades before CRISPR-Cas9. But CRISPR-Cas9 has revolutionized gene editing in a way that prior techniques have not. It has encouraged governments to critically assess, and in some cases re-think, how these techniques are regulated in research and commercialization. Today, there is a patchwork of diverse regulatory frameworks, with varying levels of acceptance of gene editing as equivalent to traditional breeding techniques. For example, Japan has commercialized two gene edited agrifoods; the GABA[2] enhanced tomato, and two types of fish for human consumption, while the United States has commercialized a soybean with improved fatty acid

---

[1] GMOs are any organism that has genetic material changed through genetic engineering that does not occur naturally by mating or natural recombination. There is diversity across jurisdictions, in terms of how a GMO is defined, but many countries base their definition on the Cartagena Protocol on Biosafety definition. A 'Living Modified Organism' (LMO) is "any living organism that possesses a novel combination of genetic material obtained through the use of modern biotechnology." Regulations for any 'novel' agrifood product are designed to protect human, animal and environmental health, though there is wide diversity in the details of how that is achieved in the regulatory space (Entine et al., 2021: 552).

[2] Gamma-aminobutyric acid (GABA) is a naturally occurring amino acid that works as a neurotransmitter. Studies have shown that GABA may help reduce anxiety, depression in humans.

composition (Waltz, 2021) and American company Pairwise is set to commercially release a gene edited leafy green.

On the research side, Chinese institutions hold a significant proportion of patents for gene edited crop varieties (Bagley & Candler, 2023), but the Chinese national government has not yet publicized any regulatory protocols for commercialization of these products in the domestic market. Canadian regulatory bodies have determined that gene edited agrifoods do not present any novel risks to human, animal or environmental health compared to conventional breeding techniques, and applications for commercialization will be assessed on a case-by-case basis for health and biosafety concerns. The EU has had legislation in place since 2001 prohibiting the deliberate release of all products of New Breeding Techniques (NBTs). Gene edited crops are GMOs under Directive 2001/18/EC. But in July of 2023, the EU Commission announced that it was re-examining how 'new genomic techniques' (including gene editing like CRISPR-Cas9) are regulated, considering a loosening of regulatory protocols for organisms that are not transgenic (no foreign genes from other species) and have less than 20 genetic modifications.

Gene editing in agrifood research, product development and regulatory frameworks is not where it was in 2018. The global economy is also not where it was in 2018. System wide shocks such as the COVID-19 global pandemic, the Russian invasion of Ukraine and its impact on global food security, and the widespread damaging effects of climate change have put enormous stresses on the agrifood system and its global supply chains. Some experts argue that we are at the tipping point and must harness useful agricultural technologies in whatever way we can to mitigate damages to human, animal and environmental health and welfare if we are to avoid catastrophic environmental collapse and widespread hunger. Prior to the discovery of CRISPR-Cas9, gene editing techniques had been around for decades, and proponents argued for their wider adoption to help manage future challenges. But the progress of gene editing as a useful and effective tool in agrifood research and development is not where some had hoped when CRISPR-Cas9 was discovered in 2012.

The status of gene editing in the agrifood system is in a state of flux. In some ways, the technology has not lived up to its potential to provide solutions to global food insecurity and the damaging effects of climate change on the world's food supply. But why? What is the relationship between the current status of gene editing applications and the governance of agrifood technologies? Where might the future take gene editing?

In this book, we investigate the factors that have guided science and governance discourse surrounding gene editing technologies, what has changed, and how might governance and regulatory frameworks respond to changes in the application of agrifood technologies in the future. The parameters of this study are as follows. We specifically focus on plants, with a brief discussion on gene edited microorganisms. Discussions of gene edited animals are not included in this study, however more general discussions of 'gene editing' techniques are analyzed in the context of crops and plant-based food & feed. Though the socio-political controversies and social movement oppositions surrounding biotechnology in agriculture are acknowledged as significant in terms of how gene editing is perceived in the public space, this book focuses on the governance of agricultural innovation of gene editing in the context of regulatory policy frameworks.[3] We discuss the influences of public perceptions of gene editing on regulatory frameworks, but our primary focus is on formalized regulations and guidelines governing the use and release of gene edited agrifoods.

Our analysis builds on previous studies of regulatory frameworks for gene editing in agrifood (see Entine et al., 2021; Ericksson et al., 2019; Menz et al., 2020; Wolt & Clark, 2018; OECD, 2018; NASEM, 2017; Friedrichs et al., 2019a; FAO, 2022; Zarate et al., 2023). Many academic and organizational studies focus on current applications of gene editing and how they are evaluated by regulatory agencies responsible for approving gene edited agrifood products for commercialization. The governance of gene edited crops has also received attention, as new applications for gene editing techniques are discovered (Phillips & Macall, 2021; Selfa et al., 2021; Asquer & Krachkovskaya, 2021; Friedrichs et al., 2019b). Methods used to investigate the dynamics of governance include interviews or surveys designed to evaluate either consumer, government, bench scientists and/or agribusiness' perceptions of regulations, risk, safety, benefits, and uncertainties regarding gene edited agrifoods (Lassoued et al., 2021).

The volume of research in this area has exploded over the last few years for three main reasons: the advancement of gene editing applications such as CRISPR to improve agronomic and nutritional traits in agrifoods; second, the evolution of regulatory guidelines in countries that have previously restricted the use of gene editing; and third, the global economic, political, and environmental turbulence of the last 5 years. The rapidly evolving global economic, environmental and security issues impacting everything from food and fuel costs, to climate, biodiversity and human migration have brought more attention. Climate change, rising global food and nutritional insecurity, global economic uncertainties, volatile commodity markets (which have contributed to spiraling food costs), and international armed conflicts are system-wide factors influencing the development of nascent regulatory frameworks. Of particular interest to contemporary academic study are those crops with gene edits that contribute to sustainability via improved yields (e.g., drought resistance), research investments in orphan crops[4] and developments in improving storability traits of staple crops (FAO, 2022: 14).

New applications may have major implications for how gene editing systems are governed in the near future, which requires ongoing evaluations by scholars, but also makes it difficult to provide an up-to-date portrait of regulatory frameworks, as they are responding to change in real-time (see Entine et al., 2021; Friedrichs et al., 2019a; Genetic Literacy Project, 2023). Most discussions of gene editing as it pertains to agrifood plants includes some discussion of regulatory frameworks. The

---

[3] For an in-depth study of the societal responses to gene editing in the agrifood system, see Selfa et al. (2021).

[4] Orphan crops are those which have significant potential to contribute to nutritional security, biodiversity and livelihoods but are grown by smallholders, often in subsistence farming systems, and thus deemed less lucrative investments for agbiotech firms.

typologies are not identically worded, but they are essentially divided into three primary categories: prohibited, regulated like GMOs, and evaluated on a case-by-case basis under existing biosafety protocols (no transgenes).

Bain et al. (2020) views the typology of regulatory regimes as based on 'three sociotechnical imaginaries', which refers to what Macnaghten et al. (2005: 279) describe as "how gene editing is being imagined by proponents and the implicit assumptions, values and visions that they assert". The three sociotechnical imaginaries pertaining to gene edited agrifoods are as follows: (1) equivalent to traditional plant breeding; (2) have the potential to usher in a new Green Revolution; and (3) could facilitate the democratization of agricultural biotechnologies. This typology focuses on the ideological underpinnings of the content of regulations governing gene edited agrifoods. Studies interviewing stakeholders have revealed important information about the perceptions of the safety of gene edited agrifoods and concerns about regulations, resource sharing and communication (see Selfa et al., 2021; Zarate et al., 2023). All three of these aspects are addressed throughout this book using several approaches and methods.

1.2 Approaches and Methods

We take a multidisciplinary systems approach, drawing insights from the natural and social sciences, including biology (agricultural science), political science (policy and institutional analyses), and agricultural economics (innovation studies, supply chain analysis). This issue area is multidisciplinary by nature, and as such, much of the current literature integrates information from scientific research on gene editing and plant breeding with perspectives from agricultural economics, public policy analysis, and/or supply chain analysis. Our analysis contributes to the recent literature examining the governance elements of regulating gene edited foods and stakeholder interactions within the broader agrifood system (Bain et al., 2020; Bogdanove et al., 2018; Eriksson et al., 2019; Gordon et al., 2021; Lassoued et al., 2021; Macnaghten et al., 2005; Philips & Macall, 2021; Schiemann et al., 2021; Sprink et al., 2020; Wolt & Clark, 2018).

The literature pays close attention to how some segments of civil society perceive gene edited foods as sharing the same risk profile as genetically modified foods, and how these risk perceptions have found their way into policy frameworks that do not necessarily reflect evidence from science-based research. Academic attention has also focused on exploring the ethical and social justice dimensions of governing gene edited agrifoods concerning who owns (or has access to) genetic resources and how scientization, privatization, and democratization of knowledge of gene-edited foods is playing out (Selfa et al., 2021; Kinchy, 2012).[5]

---

[5] The democratization of knowledge refers to the proliferation of knowledge across the wider population beyond experts and elites. Public access to information via the Internet and libraries facilitates this effort.

This book employs aspects from the multilevel perspective and the agricultural innovation systems approaches to analyze the governance dynamics of gene edited agrifoods, as part of the broader agrifood system. The multilevel perspective analyzes the dynamics between governance structures and relevant stakeholders (international organizations, national regulatory frameworks, regulators, scientific research institutions) within the global agrifood system. The multilevel perspective consists of institutional structures of the agrifood system (environmental, socio-political, and economic). We draw from innovation system studies which views these institutional structures as 'part of the background' creating opportunities for innovation processes requiring input from society (de Boon et al., 2022: 407). The broader institutional context changes, albeit slowly. Environmental, socio-political, economic, or scientific shocks to the system lead to more punctuated change. Individual stakeholders do not often influence how the system changes, but innovations can alter structures over time (Geels & Schot, 2007; Klerkx et al., 2012). Interactions and dynamics between levels of decision-making (governance), as well as dynamic interactions among innovations, socioeconomics, and environmental factors, for example, can also facilitate change within the system (Geels, 2019; Smith et al., 2010).

The Agricultural Innovation Systems (AIS) approach perceives agricultural innovation as technological change but considers institutional change as a component of innovation adoption. This approach is nestled in the complex adaptive systems approach as it emphasizes interactions between heterogenous stakeholders related to multiple dimensions of agricultural innovations, including technology development, institutional change, supply chain organization, market development, and nurturing societal acceptance of the innovation (Ekboir, 2003; Spielman et al., 2009; Klerkx et al., 2010). These are defined as self-organizing systems "whose properties cannot be analyzed by studying its components separately[…] formed by many agents of different types, where each defines his/her strategy, reacts to the actions of other agents and to changes in the environment, and tries to modify the environment in ways that fit his/her goals" (Spielman et al., 2009: 400). AIS also attends to networks of actors and influential institutional structures that condition how stakeholders interact. It allows for the ability to identify network configurations of actors and their socio-institutional context determining if the institutional context hampers or supports the innovation under study (Klerkx et al., 2010, 2012; de Boon et al., 2022).

International Regulation of Gene Editing Technologies in Crops: Current status and future trends fits in the innovation systems literature. It focuses on explaining innovation pathways in relation to public policy making (Simon, 1962; Rogers, 1995; Amaral & Ottino, 2004). Together, the multilevel perspective and the AIS view innovation as technological change, but also consider how institutional change factors into adoption. We also draw from and contribute to the innovation systems literature (de Boon et al., 2022: Klerkx et al., 2012; Geels, 2019; Geels & Schot, 2007; Smith et al., 2010). Our approach is nestled in the complex adaptive systems approach as it emphasizes interactions between heterogenous stakeholders related to multiple dimensions of agricultural innovations, including technological development, institutional change, and market development (Ekboir, 2003; Spielman et al., 2009; Klerkx et al., 2010). As a branch of network analysis, AIS attends to networks of actors and influential institutional structures that condition how stakeholders interact. It allows for the ability to identify network configurations of actors and their socio-institutional context determining if the institutional context hampers or supports the innovation under study (Resnick et al., 2015).

An important concept used in this book that is both a component of multilevel perspective and AIS is 'governance'. Governance is a horizontal approach to decision-making that includes public and private actors and the fluid dynamic between legitimacy and liability (Giddens, 1990; Majone, 1995; Moore, 2002; Kooiman, 2003). Politically, the era of governance is signified by the rise of the 'regulatory state' that is made up of 'networks of flexibility' involving both public and private actors in decision-making processes. Multiple and diverse actors involved in various stages of regulating risk make it more difficult to assign liability and blame when there is a case of regulatory failure (Moore, 2002: 122).

Academic attention has also been captured by the ethical and social justice dimensions of the governance of gene edited agrifoods concerning who owns (or has access to) genetic resources and how scientization, privatization and democratization of knowledge regarding gene edited foods is playing out (Selfa et al., 2021; Kinchy, 2012; Munawar et al., 2024). One newer area of interest is analyzing how the information economy may be influencing public debates regarding gene editing, which may have implications for how these agrifoods are regulated in the present and future. Our analysis sits within the risk governance literature, especially as it pertains to manufactured risks, of which agricultural biotechnology is considered to belong (Beck, 1992; Renn, 2008; Giddens, 1990). The deliberative element to governance—an important component of our analysis—is also derived from the literature on governance (Jasanoff et al., 2015; Macnaghten et al., 2021; Hajer, 2003). Hendriks (2009: 175) provides a succinct understanding of deliberative governance. Deliberative governance is based on the idea that, "policy making requires spaces where different institutions, agencies, groups, activists and individual citizens can come together to deliberate on pressing social issues."

We employ qualitative analysis using primary (policy) and secondary (academic articles) documents and expert interviews (investigative triangulation). Other studies on the governance of gene editing in agriculture have employed similar methods (see Selfa et al., 2021; Zarate et al., 2023). Most recently, Ruder and Kandlikar's (2023) study includes interviews with gene edited agrifoods experts in Canada and argues that there is resistance to deliberative elements of governance models. We conducted a series of in-depth interviews with regulatory experts and research scientists in various countries in summer 2023.[6] We interviewed public sector scientists and scientists in the non-profit sector, agricultural economists, regulators involved in policy discussions surrounding gene edited agrifoods, Non-Governmental

---

[6] Further details of the interview methodology and a master list of interview questions are provided in Appendix A.

Organization (NGO) representatives engaged in international agricultural development projects, and private sector representatives. Each interviewee had expertise and knowledge regarding the governance of gene edited agrifoods and/or gene editing technologies. The qualitative interview material expresses timely views on the state of gene editing in the agrifood system. Considering how quickly some aspects of the governance of gene edited agrifoods is moving, collecting expert opinions on current regulatory change and/or recent advancements in gene editing techniques was essential to gain a better understanding of the current governance dynamics, and also the future outlook for applications and regulations. Interviews with regulatory experts and research scientists who have practical experience in using gene editing technologies provide a rich set of insights which informs the policy recommendations we present in Chap. 6. Our analysis looks forward to future potential applications of gene editing technologies in the agrifood system and provides insights into how novel technologies on the horizon may be regulated in the future.

1.3   Structure of the Book

The manuscript is organized around probing two questions. Part I (Chaps. 2 and 3) seeks to answer the question: *why are there so few gene edited agrifoods on the market despite the initial optimism that accompanied the Nobel Prize-winning discovery of CRISPR-Cas9 over a decade ago?* It looks at the current suite of genomic techniques used in crop breeding to improve agronomic traits and nutritional profiles. Then it examines how current regulations governing the use and commercial release of gene edited agrifoods influence their adoption and commercialization. Part II (Chaps. 4 and 5) seeks to answer: *what governance challenges and opportunities will shape the future applications of gene editing in the agrifood system?* This part of the book explores 'new genomic techniques' and applications on the horizon. It also discusses how international regulatory frameworks can better respond to future genomic advancements in plant breeding by employing elements of deliberative governance. Deliberative approaches put effective, inclusive, and transformative communication at the heart of global governance.

In Part I—'Balancing Innovation: Application and Regulation'—we discuss current dynamics of gene editing techniques and regulation frameworks. Chapter 2, 'What is gene editing?', begins with the basics. It explains what gene editing is, what it is not, and how different techniques are used to achieve edits in plant genomes. The chapter maps out a timeline of the development of gene editing techniques, discussing the benefits and challenges of each, with most recently CRISPR-Cas9.

We then turn our attention to the current regulatory frameworks that govern the use and commercialization of gene edited agrifoods in Chap. 3 (How are gene editing technologies regulated in the agrifood system?). This chapter examines current governance and regulatory frameworks for gene edited agrifoods around the world. It details the risk assessment regimes of countries with biosafety regulatory frameworks, highlighting the similarities and differences. It discusses how effective governance of gene edited agrifoods requires balancing innovation with considerations of risk and benefits to the economy, society, and the environment. It also discusses intellectual property and 'freedom to operate' issues that arose with the proliferation of CRISPR-Cas9, and how the licensing and patent landscape has prompted some agricultural scientists to look beyond CRISPR-Cas9. In this chapter, we explore Canada's response to regulating gene edited agrifoods as a case study in reflexive governance for innovative agricultural technologies. The chapter discusses recent system-wide economic, ecological, and geo-political shocks that have influenced some countries to rethink their position on gene editing in the agrifood system.

In Part II—'Emerging Opportunities for Regulatory Enhancement'—we focus on emergent techniques and policy options for regulators as gene editing in agrifood continues to evolve, and new techniques emerge. Chapter 4 examines new breeding techniques (NBTs) and applications on the horizon and their applications for plants in the agrifood system. We discuss how NBTs can enable other technologies and platforms and highlight new classes of gene edited products in the pipeline. Finally, it examines how the regulatory trajectory of gene editing offers clues as to how new applications of NBTs may be assessed in the future as ecological pressures continue to stress the agrifood system.

In Chap. 5 (Governing the unknown: Regulating future technologies), we delve into the governance mechanisms for emerging breeding techniques. This chapter examines how governance structures and regulatory frameworks might respond to emergent new breeding techniques. The chapter discusses future policy options for assessing the biosafety and efficacy of NBTs. We argue that the primary determinant of the chosen pathway is rooted in how the regulatory systems assess other biotechnologies (precautionary, case-by-case, or restricted). The chapter finally explores how emergent techniques may pose challenges for regulatory frameworks and offers potential regulatory responses.

We conclude this book with a chapter examining next steps and future directions for studies in governance of gene editing and other NBTs in agriculture (Chap. 6). This chapter summarizes the challenges of governing gene editing in the agrifood system. It assesses the current landscape of regulatory frameworks and reviews what may change in the coming years and decades as climate change and food insecurity continue to stress global agrifood supply chains and the system writ large.

References

Amaral, L. A. N., & Ottino, J. M. (2004). Complex networks: Augmenting the framework for the study of complex systems. *European Physical Journal, B38*, 147–162.

Asquer, A., & Krachkovskaya, A. (2021). Uncertainty, institutions and regulatory responses to emerging technologies: CRISPR gene editing in the US and the EU (2012–2019). *Regulation and Governance, 15*, 1111–1127.

Bagley, M., & Candler, A. G. (2023). CRISPR patent and licensing policy. In *Assessment of the regulatory and institutional frameworks for agricultural gene editing via CRISPR-based technologies in Latin America and the Caribbean*. Genetic Engineering and Society Center. NC State University. IDB. https://doi.org/10.18235/0004904

Bain, C., Lindberg, S., & Selfa, T. (2020). Emerging sociotechnical imaginaries for gene edited crops for foods in the United States: Implications for governance. *Agriculture and Human Values, 37*, 265–279. https://doi.org/10.1007/s10460-019-09980-9

Beck, U. (1992). *Risk society: Towards a new modernity*. Sage.

Bogdanove, A., Donovan, D., Elorriaga, E., Kuzma, J., Pauwels, K., Strauus, S., & Voytas, D. (2018). Genome editing in agriculture: Methods, applications and governance. *CAST Issue Paper, 60*, 25–42.

de Boon, A., Sandstrom, C., & Rose, D. C. (2022). Governing agricultural innovation: a comprehensive framework to underpin sustainable transitions. *Journal of Rural Studies, 89*, 407–422.

Ekboir, J. M. (2003). Research and technology policies in innovation systems: Ero tillage in Brazil. *Research Policy, 32*(4), 573–586.

Entine, J., Felipe, M. S. S., Groenewald, J. H., et al. (2021). Regulatory approaches for genome edited agricultural plants in select countries and jurisdictions around the world. *Transgenic Research, 30*, 551–584. https://doi.org/10.1007/s11248-021-00257-8

Eriksson, D., Kershen, D., Nepomuceno, A., Pogson, B., Prieto, H., Purnhagen, K., Smyth, S., Wesseler, J., & Whelan, A. (2019). A comparison of the EU regulatory approach to directed mutagenesis with that of other jurisdictions, consequences for international trade and potential steps forward. *New Phytologist, 222*(4), 1673–1684.

FAO (Food and Agriculture Organisation). (2022). *Gene editing and agrifood systems*. FAO (Food and Agriculture Organisation). Retrieved January 3, 2024, from https://doi.org/10.4060/cc3579en

Friedrichs, S., Takasu, Y., Kearns, P., Dagallier, B., Oshima, R., Schofield, J., & Moreddu, C. (2019a). An overview of regulatory approaches to genome editing in agriculture (conference report). *Biotechnology Research and Innovation, 3*, 208–220.

Friedrichs, S., Takasu, Y., Kearns, P., Dagallier, B., Oshima, R., Schofield, J., & Moreddu, C. (2019b). Policy Considerations Regarding Genome Editing. *Trends in Biotechnology, 37*(10), 1029–1032.

Geels, F. W. (2019). Socio-technical transitions to sustainability: A review of criticism and elaborations of the Multi-Level Perspective. *Current Opinion in Environmental Sustainability, 39*, 187–201.

Geels, F. W., & Schot, J. (2007). Typologies of sociotechnical transition pathways. *Research Policy, 36*(3), 399–417.

Genetic Literacy Project. (2023). *Gene editing regulation tracker*. Retrieved December 13, 2023, from https://crispr-gene-editing-regs-tracker.geneticliteracyproject.org/

Giddens, A. (1990). *The consequences of modernity*. Polity Press.

Gordon, D. R., Jaffe, G., Doane, M., Glaser, A., Gremillion, T. M., & Ho, M. D. (2021). Responsible governance of gene editing in agriculture and the environment. *Nature Biotechnology, 39*, 1055–1057. https://doi.org/10.1038/s41587-021-01023-1

Hajer, M. (2003). A frame in the fields: Policy making and the reinvention of politics. In M. Hajer & H. Wagenaar (Eds.), *Deliberative policy analysis: Understanding governance in the network society* (pp. 88–110). Cambridge University Press.

Hendriks, C. (2009). Deliberative governance in the context of power. *Policy and Society, 28*(3), 173–184.

Jasanoff, S., Benjamin Hurlbut, J., & Saha, K. (2015). CRISPR democracy: Gene editing and the need for inclusive deliberation. *Issues in Science and Technology, 32*(1), 25–32.

Kinchy, A. (2012). *Seeds, science and struggle: The global politics of transgenic crops*. MIT Press.

Klerkx, L., Aarts, N., & Leeuwis, C. (2010). Adaptive management in agricultural innovation systems: The interactions between innovation networks and their environment. *Agricultural Systems, 103*(6), 390–400.

Klerkx, L., van Mierlo, B., & Leeuwis, C. (2012). Evolution of systems approaches to agricultural innovation: concepts, analysis and interventions. In I. Darnhofer, D. Gibbon, & B. Dedieu (Eds.), *Farming systems research into the 21st century: the new dynamic* (pp. 457–483). Springer.

Kooiman, J. (2003). *Governing as governance*. Sage.

Lassoued, R., Phillips, P. W. B., Macall, D. M., Hesseln, H., & Smyth, S. J. (2021). Expert opinions on the regulation of plant genome editing. *Plant Biotechnology Journal, 14*(4), 321–337. https://doi.org/10.1111/pbi.13597

Macnaghten, P., Kearnes, M. B., & Wynne, B. (2005). Nanotechnology, governance, and public deliberation: What role for the social sciences? *Science Communication, 27*(2), 268–291.

Macnaghten, P., Shah, E., & Ludwig, D. (2021). Making dialogue work: Responsible innovation and gene editing. In D. Ludwig, B. Boorgaard, P. Macnaghten, & C. Leeuwis (Eds.), *The politics of knowledge in inclusive development and innovation* (1st ed., pp. 243–255). Routledge.

Majone, G. (1995). The rise of the regulatory state in Europe. *West European Politics, 17*, 77–101.

Menz, J., Modrzejewski, D., Hartung, F., Wilhelm, R., & Sprink, T. (2020). Genome edited crops touch the market: A view on the global development and regulatory environment. *Frontiers in Plant Science, 11*, 586027. https://doi.org/10.3389/fpls.2020.586027

Moore, E. (2002). The new direction of federal agricultural research in Canada: From public goods to private gain? *Journal of Canadian Studies, 37*, 112–134.

Munawar, N., Ahsan, K., & Ahmad, A. (2024). Chapter 18: CRISPR-edited plants' social, ethical, policy, and governance issues. In K. A. Abd-Elsalam & A. Ahmad (Eds.), *Global regulatory outlook for CRISPRized plants* (pp. 367–396). Academic. https://doi.org/10.1016/B978-0-443-18444-4.00011-9

NASEM. (2017). *Preparing for future products of biotechnology*. NASEM. https://doi.org/10.17226/24605, https://www.ncbi.nlm.nih.gov/books/NBK442207/

Organisation for Economic Co-operation and Development (OECD). (2018). *Conference on genome editing: Applications in agriculture*. Retrieved on December 2, 2023, from: https://www.oecd.org/environment/genome-editing-agriculture/

Phillips, P. W. B., & Macall, D. (2021). *Environmental scan of common practices in genome editing and CRISPR in Canadian public research institutions* (Working Paper Series (2021-03)). Centre for the Study of Science and Innovation Policy (CSIP). https://www.schoolofpublicpolicy.sk.ca/csip/documents/research-paper-summaries/2021_environmental-scan-of-common-practices-in-genome-editing-and-crispr-in-canadian-public-research-institutions.pdf

Renn, O. (2008). Risk governance: Combined facts and values in risk management. In J. H. Bischoff (Ed.), *Risks in modern society: Topics in safety, risk, reliability, and quality* (pp. 61–118). Springer.

Resnick, D., Babu, S.C., Haggblade, S., Hendriks, S. & Mather, D. (2015). *Conceptualizing drivers of policy change in agriculture, nutrition, and food Security: The Kaleidoscope model*. IFPRI Discussion Paper 01414. Retrieved September 13, 2023, from: https://ssrn.com/abstract=2564542

Rogers, E. M. (1995). *Diffusion of innovations* (4th ed.). The Free Press.

Ruder, S. L., & Kandlikar, M. (2023). Governing gene-edited crops: risks, regulations, and responsibilities as perceived by agricultural genomics experts in Canada. *Journal of Responsible Innovation, 10*(1), 2167572. https://doi.org/10.1080/23299460.2023.2167572

Schiemann, J., Hartung, F., Menz, J., Sprink, T., & Wilhelm, R. (2021). Policies and governance for plant genome editing. In A. Ricroch, S. Chopra, & M. Kuntz (Eds.), *Plant biotechnology*. Springer. https://doi.org/10.1007/978-3-030-68345-0_18

Selfa, T., Lindberg, S., & Bain, C. (2021). Governing gene editing in agriculture and food in the United States: Tensions, contestations, and realignments. *Elementa: Science of the Anthropocene, 9*(1), 00153. https://doi.org/10.1525/elementa.2020.00153

Simon, H. A. (1962). The architecture of complexity. *Proceedings of the American Philosophical Society, 106*, 467–482.

Smith, A., Voß, J.-P., & Grin, J. (2010). Innovation studies and sustainability transitions: The allure of the multi-level perspective and its challenges. *Research Policy, 39*(4), 435–448.

Spielman, D. J., Ekboir, J., & Davis, K. (2009). The art and science of innovation systems inquiry: Applications to Sub-Saharan African agriculture. *Technology in Society, 31*(4), 399–405.

Sprink, T., Wilhelm, R. A., Spök, A., Robienski, J., Schleissing, S., & Schiemann, J. H. (Eds.). (2020). *Plant genome editing—Policies and governance*. Frontiers Research Topics. Frontiers Media. https://doi.org/10.3389/978-2-88963-670-9

Waltz, E. (2021, December 14). GABA-enriched tomato is first CRISPR edited food to enter market. *Nature Biotechnology, 40*(1), 9–11. https://doi.org/10.1038/d41587-021-00026-2

Wolt, J. D., & Clark, W. (2018). Policy and governance perspectives for regulation of genome edited crops in the United States. *Frontiers in Plant Science, 9*, 606. https://www.frontiersin.org/articles/10.3389/fpls.2018.01606

Zarate, S., Cimadori, I., Mercedes Roca, M., Jones, M. S., & Barnhill-Dilling, K. (2023). *Stakeholder interviews assessment of the regulatory and institutional framework for agricultural gene editing via CRISPR-based technologies in Latin America and the Caribbean*. Genetic Engineering and Society Center, NC State University. Retrieved on July 12, 2023, from: https://research.ncsu.edu/ges/files/2023/01/IDB-Crispr_Stakeholder-Interviews_2023.pdf

Chapter 2
What Is Gene Editing?

I would be more worried about pesticides and chemicals that are used than about a simple mutation, which if allowed enough time and enough resources might exist anyways.— Informant 18 (research scientist/NGO representative)

2.1 Introduction

The concept of editing a genome with the aid of a programmable nuclease emerged as part of advancements in functional genomics over 20 years ago (National Academies of Sciences & Medicine (NASEM), 2020). This field is focused on understanding the relationship between the information contained in an organism's genome and its physical characteristics. All genome editing techniques rely on the single step of engineering an enzyme (i.e., the nuclease), that induces a Double-Strand Break (DSB) at a specific site of the DNA that is to be edited.[1] Unlike genetic modification, gene editing does not use foreign nucleotides to induce change in DNA. Instead, it harnesses natural repair processes found within the cell.

In almost every discussion of gene editing in agrifood, a description of various gene editing techniques is included. Gao's (2018) piece in the scientific journal *Cell* compares gene-editing techniques in detail, as does the FAO's comprehensive *Gene Editing and Agrifood Systems* document (FAO, 2022). Explaining the details of different techniques used to perform gene editing in language accessible to non-experts provides readers with accurate information about what gene editing is, what it is not and what it can and cannot do.[2] This chapter attempts to do just that. Briefly, explain gene editing techniques to better understand the dynamics of governance and regulation surrounding this groundbreaking technology as it is used in agrifood systems.

---

[1] For terms related to genome editing techniques, please refer to "Glossary."

[2] Another way that experts in this area have helped to make discussions surrounding gene editing more accessible is through published glossaries of scientific terms that are often used in discussions. The National Academies of Sciences, Engineering and Medicine (NASEM) (2017b) published a detailed glossary of terms explaining what specific words mean in accessible language. Though this specific document focuses on human genome editing, much of the terminology is similar to terms used in agribiotech discussions.

L. F. Clark, J. E. Hobbs, *International Regulation of Gene Editing Technologies in Crops*, SpringerBriefs in Environmental Science, https://doi.org/10.1007/978-3-031-63917-3_2

2.2   New Breeding Techniques

The most advanced New Breeding Techniques (NBTs) alter an organism's genome by using gene editing and can be used in several different ways. They can be used to replace a disease-causing mutation sequence with a normal sequence. NBTs can also be used to disrupt an expressed gene by turning it off.[3] The genome is altered by targeting nucleotides via types of variants. This can be done by using one of two techniques: Oligo-Directed Mutagenesis (ODM) or Site-Directed Nucleases (SDN) (Lassoued et al., 2021). ODM is a rapid, precise, non-transgenic plant breeding alternative that uses synthetic oligonucleotides[4] that are like DNA molecules with the target sequence (homologous) but for the nucleotide(s) to be modified. Oligonucleotides target the homologous sequence and create a mismatch at the base pair that is to be modified. This mismatch is recognized by the DNA repair machinery of the cell and the mismatch is repaired introducing the altered nucleotide.

2.2.1   Site Directed Nucleases

There are three types of Site-Directed Nucleases (SDN) (see Table 2.1). SDN uses DNA-cutting enzymes (nucleases) that are instructed to cut the DNA at a specific location by several binding systems. After the cut is made, the cell's own DNA repair mechanisms recognize the break and repair it using one of two cell repair processes. In the case of the Non-Homologous End-Joining (NHEJ) pathway, there is no donor DNA. The cut in DNA is rejoined, however this may cause a few base pairs to be eaten away or added, resulting in random, small deletions or additions (a few base pairs) of nucleotides at the cut site (SDN1) (Entine et al., 2021: 559). The Homology-Directed Repair (HDR) pathway involves a donor DNA that carries the chosen change and has homology[5] with the target site used to introduce the chosen change at the cut site. This allows for an introduction of specific intentional insertions, changes, or deletions. The SDN2 technique targets a gene for correction. The SDN3 technique inserts a gene into the DNA.

---

[3] Gene expression is the process by which RNA and proteins are made from instructions in the genes. Alterations in gene expression change the functions of cells, tissues, organs, or whole organisms and sometimes result in observable characteristics associated with a particular gene, such as eye colour or hair colour (adapted from NASEM, 2017b: 264).

[4] Synthetic oligonucleotides are short nucleic acid chains that can act in a specific manner to control gene expression.

[5] Homology refers to a sequence of two different genes that are similar and emerged from a common evolutionary ancestor gene.

Table 2.1 Three main types of gene editing using Site-Directed Nucleases (SDN)

	Purpose	Example
SDN1	Involves unguided repair of targeted double-strand break (DSB) by nonhomologous end joining (NHEJ). Spontaneous repair of DSB can lead to mutations causing gene silencing, gene knockout or changes in gene activity. No exogenous DNA repair template is used in these applications	Disease resistance in banana, cassava, maize, rice
SDN2	The objective is to generate a localized pre-defined point mutation or deletion/addition, due to co-introduction of a repair DNA molecule that is homologous to the targeted area and is expected to act as a repair template. Repairing is achieved by recombining two like DNA molecules (homologous recombination (HR)). SDN2 generates changes spanning few base pairs in genetic elements (promoters, coding sequences, etc.) that pre-exist in the host genome. Efficiency is lower than SDN1, but varies according to species, donor design, time and method of delivery, and other conditions	Disease resistance in potato and wheat; insect resistance in rice
SDN3	The objective is to generate a localized pre-defined insertion/ deletion/replacement of entire genetic elements (promoters, coding sequences, etc.), or entire genes, because of co-introducing a large DNA molecule to be inserted in the target area. DNA molecule may or may not be homologous to the targeted area, and its insertion can take place either by HR or by NHEJ. Involves template guided repair of targeted DSB using a sequence donor, typically double-stranded DNA containing entire gene or even longer genetic element(s); both ends of donor are homologous. Efficiency is lower than SDN1, but varies according to species, donor design, time and method of delivery and other conditions	Herbicide tolerant soybean; drought resistant maize

Adapted from FAO (2022), Friedrichs et al. (2019)

In some cases, countries regulate the three main types of gene editing differently. Since the SDN3 technique sometimes uses transgenes to edit the genome (transgenics), its risk profile is treated differently from gene editing that uses SDN1 or SDN2 in some contexts. Some argue that mutations in a controlled environment regardless of the technology used may, in fact, be safer than those occurring in nature. As research scientist Informant 5 put it,

> I find that technologies tend to be exceptionalized in the food system. And the reality is that technologies are probably the least risky part of how we produce and consume food.

We asked Informant 17, who is a research scientist, how they view gene editing compared to mutagenesis in terms of their safety profiles. When asked what they wish laypersons knew about gene editing, Informant 17 responded,

> …that it is equivalent to traditional mutagenesis approach. For adapting gene function. And that it's safer than those mutagenesis approaches. Because when you apply mutations, you get a plethora of mutants in every individual plant and then you screen them for something of interest related to your gene of interest. And then you have to back cross many, many times to get rid of all the background mutations.

Table 2.2 Gene editing techniques

Technique	Purpose
Meganucleases	A special type of enzyme that binds to/cuts DNA at specific sequences of a length that occurs at few sites in the genome. These are natural enzymes (and their synthetic derivatives) that catalyze DNA rearrangement events via DNA cleavage
Zinc Finger Nucleases (ZFN)	Artificial proteins (engineered enzymes) consisting of a nuclease domain (usually Fokl) coupled to a zinc finger domain. Proteins interact with three specified base pairs of DNA causing double-strand breaks on the targeted site in the genome
TALENs	Artificial protein complexes with unique DNA-binding domain in which a nuclease is coupled to a Transcription Activator-Like (TAL) effector domain, which is capable of recognizing and binding specific genetic sequences. TAL effectors can induce site-specific mutations in a genome. Can be engineered to bind to any DNA sequence
CRISPR-Cas9	A component of the adaptive immunity system in bacteria; uses site-specific nucleases (SSNs) to generate double-strand breaks in specific genes at targeted locations in the genome. CRISPR-Cas9 nucleases are guided to a certain genomic DNA sequence by guide RNAs attached to the nuclease

Adapted from NASEM (2016, 2017a), Seyran and Craig (2018), Lassoued et al. (2018), Friedrichs et al. (2019), Gatica-Arias (2020)

Fig. 2.1 Timeline of gene editing techniques. (Source: Adapted from Tröder & Zevnik, 2022)

Informant 16, also a research scientist, concurred with the above comments, stating,

> I think the general idea would be that we are not doing anything new that nature hasn't been done before. And the reason behind that is that the more media find out about genomes of the organisms around us, the more we discover that they have been moving pieces of their DNA all around throughout evolution history. We can find fragments of raw genome, we can find fragments of viral genome in the human genome.

This point is not often included in contemporary discussions about the safety and efficacy of gene editing. In many jurisdictions, gene editing and transgenics are regulated differently than techniques used in conventional breeding like mutagenesis. We discuss how these techniques are regulated in Chap. 3.

Techniques used in plant breeding aim to achieve intentional and precise knock-outs, or a (re)introduction of a desired trait. These techniques include Meganucleases, Zinc Finger Nuclease (ZFN), TAL Effector Nucleases (TALEN) and CRISPR-Cas9 (PRRI, 2023) (Table 2.2).

Figure 2.1 shows the published discoveries and descriptions of gene editing techniques from 1985 until 2020. The following section discusses the four main techniques that continue to be used to this day.

2.2.2 Meganucleases

The study of meganucleases in 1985 first revealed the basic mechanisms of DNA cleavage and the DNA repair processes on which genome editing depends. Meganucleases are a special type of enzyme that binds to and cuts DNA at specific sequences that occur at a few sites in the genome. Meganucleases are single proteins that recognize a sequence in the DNA and break the target DNA, leaving a double-strand break that can be repaired through a natural repair process used to repair broken DNA (Silva et al., 2011). They naturally occur in bacteria, single celled organisms, plants, animals, and fungi. Meganucleases can be used in genome editing for both nonhomologous end joining and homology directed repair–mediated alterations. Meganucleases-mediated genome editing has been demonstrated in maize (*Zea mays*) and tobacco (*Nicotiana* spp.) (Baltes & Voytas, 2014). Production is difficult, and cost of production is costly. It can take months to conduct an experiment using meganucleases. It is difficult to change the target sequence specificity of meganucleases, so they are not widely used for genome editing (NASEM, 2016).

2.2.3 Zinc Finger Nuclease

One of the first reliable, successful methods of genome editing was reported in 2003, though it was discovered in 1985. Zinc Finger Nuclease (ZFN) interacts with three specific base pairs of DNA that cause double-strand breaks at a targeted site in the genome. It requires two proteins. ZFNs are used to introduce mutations via Non Homologous End Joining (NHEJ), which is a natural repair process used to join the two ends of a broken DNA strand back together. ZFNs have been used to modify plants, but the technique is not always accurate as it sometimes targets the wrong sequence (NASEM, 2016). An example of where this technique has been used is in the modification of the endogenous tobacco *acetolactate synthase* genes, which is the target enzyme for two types of herbicides. By using ZFN and a donor molecule, mutations were induced, thus generating plants which were herbicide resistant (see Novak, 2019). The production and cost of production of ZFNs is prohibitively expensive. It can take months to conduct an experiment using ZFNs for gene editing.

2.2.4 TALENs

Transcription activator-like effectors (TALEs) followed zinc finger nucleases and preceded CRISPR-Cas9 as effective genome editing tools. The discovery of TALEs was first published in 2009 in *Science* by scientists at Martin Luther University in Halle, Germany. TALEs are proteins with a unique DNA-binding domain that presents a predictable and programmable specificity. TALEs are made and used by plant pathogenic bacteria to control plant genes during infection. In nature, TALEs bind to plant DNA sequences and activate genes. The bacteria encode TALEs through a simple code that has been exploited to engineer proteins with custom site specificity in any target genome (Boch et al., 2009; Moscou & Bogdanove 2009; NASEM, 2016, 2017a). The ease of the design for specific target DNA sequences of the TALEs revolutionized genome editing.

Like ZFNs, TALEs can be fused with the nuclease domain of FokI (a restriction endonuclease—a 'cleaver' enzyme) and utilized to edit the genome, referred to as Transcription Activator-Like Effectors Nucleases (TALENS). TALENs are used in pairs like ZFNs to affect targeted mutations. TALENs have a higher efficiency than ZFNs and have been used to alter the genomes of a variety of different organisms. TALENs have been used to edit genomes in rice, maize, wheat, and soybean (Baltes & Voytas, 2014; NASEM, 2016). TALENs have also been used experimentally to correct mutations that cause human disease. Like ZFNs, TALENs experiments require two proteins. Production is relatively easy, and significantly more cost effective than meganucleases and ZFNs. Gene editing experiments using TALENS can take weeks, as opposed to months as with previous techniques. However, researchers continue to improve upon TALENS. In 2020, a rapid TALENs preparation protocol was developed. This has improved reproducibility and efficiency of this gene editing technique according to the International Service for the Acquisition of Agribiotech Applications (ISAAA, 2023).[6]

TALENs has been used in several plants to improve agronomic characteristics or nutritional profiles. For example, TALENs has been used to create soybeans with low levels of polyunsaturated fats. Oils with lower levels of these fats are considered healthier compared to oils that can be hydrogenated to produce trans-fatty acids (ISAAA, 2023). Soybeans that produce high oleic acid (the 'healthier' fat) emerged on the US market beginning in 2019 in the form of soybean oil. TALENs has also been used in rice to breed resistance to bacterial blight. Through gene editing, scientists were able to generate inheritable disease resistance. TALENs has also been used to reduce acrylamide in potatoes, maize, and wheat, as well as breed resistance to wheat powdery mildew. Further applications of TALENs include utilizing sugarcane and algae in the biofuels industry (ISAAA, 2023).

2.2.5   CRISPR-Cas9

The *Clustered Regularly Interspaced Short Palindromic Repeats* (CRISPR-Cas9) system as a gene editing tool was reported in 2012 (Doudna & Charpentier, 2014). CRISPR is a naturally occurring mechanism found in bacteria. Bacteria harbour

---

[6]See ISAAA (2023) Pocket K No. 59: Plant Breeding Innovation: TALENs Transcription Activator-Like Effector Nucleases. https://www.isaaa.org/resources/publications/pocketk/59/default.asp

CRISPR as an innate defense mechanism against viruses and plasmids[7] that uses RNA-guided nucleases to target the break or cut of foreign DNA sequences. The bacteria retain fragments of foreign DNA which provides it with some immunity to viruses. CRISPR-Cas systems can generate a range of DNA edits which are synonymous with those found in natural populations. Multiple genetic changes can be achieved in a single generation (Lyzenga et al., 2021). CRISPR-Cas9 has received the most academic attention, namely because it is more precise in targeting specific genes than other techniques and is more economical to use than other gene editing techniques mentioned above. This has improved the timeliness of experimenting with this technique, and reduced the costs associated with gene editing research. It is the most recently introduced of all gene editing platforms and appears to have the highest accuracy of all gene editing techniques discussed here.

CRISPR-Cas9 is the gene-editing platform in which RNA homologous with the targeted gene is combined with the Cas9 (DNA snipping enzyme). Cas9 gives CRISPR the ability to alter DNA sequences. Cas9 makes up part of the "toolkit" for the CRISPR-Cas9 system of genome editing. The other is a homing device that can be programmed to target the DNA sequence of interest, imparting precise control over the location of edits. Scientists have dissected the innate CRISPR-Cas9 system and re-engineered it in such a way that a single RNA, the guide RNA, is needed for Cas9-mediated cleavage of a target sequence in a genome (Alvarez, 2021; Sander & Joung, 2014).

Guide RNA (short segments of RNA used to direct the DNA-cutting enzyme to the target location in the genome) design requirements are limited to a unique sequence of about 20 nucleotides in the genome (to prevent off-target effects) and are restricted near the Protospacer Adjacent Motif sequence (PAM),[8] which is specific for the CRISPR-Cas9 system. Newer applications of CRISPR include the use of two guide RNAs with a modified nuclease that "nicks" one strand of the DNA, providing greater specificity for targeted deletions. The ease of design, the specificity of the guide RNA, and the simplicity of the CRISPR-Cas9 system have resulted in rapid demonstration of the utility of this method of editing genomes in plants and other organisms (Baltes & Voytas, 2014).

CRISPR-Cas9 is a versatile and robust gene editing tool for crop improvement. Because of its efficiency and accuracy, this method is rapidly becoming the most widely used approach for performing gene editing. For example, researchers from the Institute of Agricultural Technology of Argentina (INTA) used CRISPR-Cas9 to develop non-browning potatoes and lactose-free cow's milk. CRISPR-Cas9 has also been used to increase alfalfa productivity and quality (Laaninen, 2020: 8).

---

[7] A plasmid is an extrachromosomal DNA molecule within a cell that is physically separated from chromosomal DNA and can replicate independently. It can be introduced via bacteria, plants, animals, fungi, or unicellular organisms.

[8] 'Protospacer adjacent motif sequence' (PAM) is a 2–6-base pair DNA sequence immediately following the DNA sequence targeted by the Cas9 nuclease in the CRISPR bacterial adaptive immune system. The PAM is a component of the invading virus or plasmid, but is not found in the bacterial host genome and hence is not a component of the bacterial CRISPR locus (Karvelis et al., 2015).

CRISPR-mediated gene knockout has been applied in rice, barley, soy, maize, wheat, tomato, potato, lettuce, citrus trees, mushrooms, cucumbers, grapes, watermelon, and others (Liang et al., 2014).

What seems to baffle some of the scientists we interviewed is the artificial distinctions arising in public debates over the safety and risks of gene editing—between genomic editing and mutations that take place in a controlled laboratory setting versus mutations that happen in nature. The distinction between manufactured risks and natural risks seems to be at the centre of the debate over the safety of gene editing in the agrifood system; though at a conceptual level, mutations or changes to genomes happen in nature. Informant 6 and Informant 7 (an NGO representative and a research scientist, respectively) had similar things to say about how the perception of the safety of gene editing compared to 'conventional' breeding techniques such as mutagenesis is distorted in the public dialogue on biotechnology in the agrifood system. Informant 7 notes,

> …if you compare [gene editing] to a conventional breeding technique like a random mutagenesis which is considered conventional, it's in the organic section of the grocery store, you get hundreds to thousands of random mutations per plant, using those techniques. And what? It's fine. Those plants have been on the market and we've been eating them for a hundred years, and they're fine. But that's a lot of unintended mutations that we don't even know about. Whereas with gene editing, if you get unintended mutations off-target first of all, we tend to know exactly where they would be, because they happen in regions that are very similar to the region that we're targeting. We can scan those, and we do. We look at that, and we can tell if it's been enough target mutation. …. And secondly, we're certainly not getting thousands from the gene editing process itself.

Informant 7 had more to say about genetic mutations in agrifood plants. They continue,

> they're fine, I mean one rice plant. It produces a seed and that seed has around 40 mutations, random. We don't know where they are. They're all unintended. They're all just…who knows? And it's fine. No problem…Mutations happen, and that's a good thing. That's how things evolve. That's how things adapt.

CRISPRoff/on

Another recent discovery in the CRISPR-Cas system is 'CRISPRoff' which is a technique using CRISPR-Cas9 to turn genes 'on', but which also has the potential to turn them 'off' (Nunez et al., 2021). CRISPRoff is a reversible tool for controlling gene expression that is specific, precise, and inheritable. It involves the addition of a chemical tag to the DNA making it inaccessible for reading and subsequent protein production. In the paper published in *Cell*, Nunez et al. (2021) describe how to modify CRISPR's basic architecture to extend its reach beyond the genome and into what is known as the *epigenome*—proteins and small molecules that latch onto DNA and control when and where genes are switched on or off. The researchers show that once a gene is switched off, it remains inert in the cell's descendants for hundreds of generations, unless it is switched back on with a complementary tool called CRISPRon.

Discussions have emerged regarding how to classify organisms that have had CRISPRoff applied to their genome. If the genome is not changed, but there is genetic manipulation, is it a GMO? Identification of organisms produced using CRISPRoff would be difficult to identify, as critics of gene editing in general have noted (Williams, 2021). These are advancements that may require new regulatory scrutiny in the coming years as their applications evolve, and the need for innovative ways to address climate change and food insecurity become even more pressing. However, many countries, such as Canada and Argentina, have designed their regulatory frameworks based on a risk assessment of products, not how the organism was developed. Therefore, as the National Academies of Sciences, Engineering and Medicine (NASEM) (2017a, p. 108) document states, "risk assessment endpoints for future biotechnology products are not new compared with those that have been identified for existing biotechnology products". The benefit of having a regulatory system tooled to conduct risk assessments for previous and current biotechnologies in the agrifood system on a case-by-case basis is the built-in ability to assess future, unknown technologies for future, unknown risks. However, regulatory systems may have to continually update and adapt to assess yet unknown risks that accompany new breeding techniques. As the NAESM (2017a) report cautions—if regulatory systems are not prepared for the wave of new breeding techniques on the horizon, there will be stifling of innovation and a slowing in the development of products that can be useful tools in combating food insecurity and climate change.

Table 2.3 lists applications of gene editing in the agrifood plant breeding space, including the type of NBT used in the breeding process, the type of quality improvement (food & feed and/or agronomic), the trait that was the specific focus of the breeding initiative, country of origin and research organization(s) responsible for developing the application.

2.3 Conclusion

Due to their accuracy, lower costs and application simplicity, genome editing tools can introduce valuable quantitative and qualitative traits into plants. Current research is focusing on improving agronomic traits including drought resistance, increased yield, pathogen resistance, and decreased time to ripening. CRISPR-Cas9 platforms are not perfect, and can result in off-target modifications, unwanted on-target modifications, and genomic rearrangements. ZFN and TALENs systems continue to evolve and are improving in terms of accuracy, cost and time requirements. A number of new techniques are on the horizon and are discussed in greater detail in Chap. 4.

Table 2.3 Applications of gene editing in agrifood plant breeding

	Type of NBT used	Type of quality improvement	Trait	Country of origin	Research organization
Staple crops					
Barley	CRISPR-Cas9	Food & feed	Starch-free endosperm	Australia, Bangladesh, China	Agriculture & Food Commonwealth Scientific and Industrial Research Organization, Canberra
Cassava	CRISPR-Cas9	Agronomic	Disease resistance; reduced cyanide	USA	University of California, Berkeley
Golden rice	CRISPR-Cas9	Agronomic	Increased vitamin A		University of California (Berkeley, Davis), California
Maize	CRISPR-Cas9	Agronomic	Disease resistance	USA	Dupont Pioneer, Iowa
Potato	TALENS; CRISPR-Cas9	Agronomic; food & feed	Disease resistant varieties; less acrylamide formation	Russia, USA	Russian Academy of Sciences; Cellectis Plant science, INC. Minnesota
Rice	CRISPR-Cas9; TALEN	Agronomic	Salt tolerance; disease resistance	India, USA, China	National Institute for Plant Biotechnology, New Delhi; Dept. of Genetics, Development &Cell Biology, Iowa State U.; Key Laboratory of Rice Genetic Breeding of Anhui Province, Rich Research Institute, Anhui Academy of Agricultural Sciences
Sorghum	CRISPR-Cas9	Agronomic; Food & feed	Disease resistance; increased protein content	Australia, Kenya	University of Queensland; Kenyatta University
Soybean	CRISPR-Cas9	Agronomic; food & feed	Nematode resistance; improved fatty acid comp.	Brazil, USA	Evogene, Rehovot, Israel & TMG; Calyxt, Minnesota
Wheat	CRISPR-Cas9; TALEN	Agronomic; food & feed	Fungus protection; low gluten content	China, Spain, Netherlands	State Key Laboratory of Plant Cell and Chromosome Engineering, Institute of Genetics and Developmental Biology, Chinese Academy of Sciences

Other crops

Alfalfa	CRISPR- Cas9	Agronomic	Increase yield	Argentina	National Institute of Agricultural Technology
Apple	CRISPR- Cas9	Food & feed	Accumulated tartaric acid (non-browning)	Japan, China, Italy, Ireland; South Korea	Faculty of Bioscience and Bioindustry, Tokushima U; Beijing Key Laboratory of Grape Sciences and Enology, Key Laboratory of Plant Resources, Institute of Botany, Chinese Academy of Sciences; Division of Fruit Breeding and Genetics, NARO Institute of Fruit Tree and Tea Science, Tsukuba, Ibaraki; NARO Institute of Fruit Tree and Tea Science, Tsukuba; Faculty of Ag, Iwate U, Morioka, Iwate; Research and Innovation Centre, Genomics and Biology of Fruit Crop Department, Fondazione Edmund Mach, San Michele all' Adige; Research Centre for Viticulture and Enology, CREA, Conegliano, TV, Italy; PLANTeDIT Pvt Ltd, Cork; ToolGen, Seoul; Italian Institute of Tech, Genova
Banana	CRISPR- Cas9	Agronomic	Disease resistance	Australia, Nigeria, South Africa	Queensland University of Tech; International Institute of Tropical Agriculture; Agricultural Research Council, Pretoria
Beet (sugar)	CRISPR- Cas9	Agronomic	Disease resistance	Russia	Russian Academy of Sciences
Cacao	CRISPR- Cas9	Agronomic	Disease resistance	USA	Pennsylvania State U.
Camelina	Mutagenesis	Food & feed	Improved fatty acid comp.	USA	Dept. of Plant Sciences and Plant Pathology, Montana State
Carrot	CRISPR- Cas9	Food & feed	Decreased anthocyanin acid content	Poland, USA	Institute of Plant Biology and Biotechnology, Faculty of Biotechnology and Horticulture, Krakow; Dept. of Biology, East Carolina U; Dept. of Plant Science and Landscape Architecture, U of Maryland; Institute for Bioscience and Biotechnology Research, U of Maryland
Cherry	RNAi (gene silencing)	Agronomic	Disease resistance	USA	Dept. of Horticulture, Plant Biotechnology Resource and Outreach Centre, Michigan State U
Citrus	CRISPR- Cas9	Agronomic	Disease resistance	China	Chinese Academy of Sciences
Cucumber	CRISPR- Cas9	Agronomic	Disease resistance	Israel	Dept. of Plant Pathology and Weed Research, ARO, Volcani Center

(continued)

Table 2.3 (continued)

	Type of NBT used	Type of quality improvement	Trait	Country of origin	Research organization
Eggplant	CRISPR- Cas9	Food & feed	Decreased browning	Italy, Spain	DISAFA, Plant Genetics and Breeding, U of Torino; Crop Biotechnology Department, Instituto de Biologia Molecular y Celular de Plantas, CSIC-UPV; Instituto de Conservacion y Mejora de la Agrodiversidad Valenciana, Universitat Politecnica de Valncia
Flax	CRISPR- Cas9; TALEN	Agronomic	Herbicide tolerance	USA	CIBUS, San Diego
Golden banana	CRISPR- Cas9	Agronomic	Disease resistance, creased beta carotene	India	National Agri-Food Biotechnology Institute (NABI), Department of Biotechnology, Ministry of Science and Technology
Golden flesh melon	CRISPR- Cas9	Agronomic	Increased beta carotene	Israel, USA	Plant Science Institute, Department of Plant Breeding and Genetics, Cornell University
Grape	CRISPR- Cas9	Food & feed	Decreased tartaric acid content	China	Beijing Key Laboratory of Grape Science and Enology and Key Laboratory of Plant Resource, Institute of Botany, Chinese Academy of Sciences; University of Chinese Academy of Sciences, Beijing; Sino-Africa Joint Research Center, Chinese Academy of Sciences, Wuhan
Grapevine	CRISPR- Cas9	Agronomic	Drought tolerance, disease resistance	South Africa	see Research Organizations for apple
Lettuce	CRISPR- Cas9	Food & feed	Increased vitamin C	China	State Key Laboratory of Plant Cell and Chromosome Engineering, Center for Genome Editing, Institute of Genetics and Developmental Biology, Chinese Academy of Sciences
Mushroom	CRISPR- Cas9	Food & feed	Non-browning	Japan	Graduate School of Ag, Kyoto U; Graduate School of Tech, Industrial and Social Sciences, Tokushima U
Oilseed rape	CRISPR- Cas9	Agronomic; food & feed	Herbicide tolerance; improved fatty acid comp.	China, Spain, Netherlands	Key Laboratory of Plant Functional Genomics nf the Min. of Edu, Yangzhou University; National Key Laboratory of Crop Genetic Improvement, Huazhong Agricultural University, Wuhan

Crop	Technique	Category	Trait	Country	Institutions
Peanut	CRISPR-Cas9	Food & feed	Increased oleic acid	USA, China	Tuskegee University, Alabama; Shandong Peanut Research Institute, Qingdao; Hainan University, Haikou; Guangxi Academy Of Ag Sciences, Nanning; Henan Academy of Ag Sciences, Zhengzhou; Auburn University, Alabama
Pomegranate	CRISPR-Cas9	Food & feed	Increased gallic acid and glucosides	China	Shanghai Key Laboratory of Plant Functional Genomics and Resources, Shanghai; Shanghai Chenshan Plant Science Research Center, Chinese Academy of Sciences; Dept. of Plant Sciences, U of California
Tomato	CRISPR-Cas9	Agronomic; food & feed	Disease resistance; D3 enhancement; high GABA content	USA, UK, Japan	Sanatech Minato-ku, Tokyo and University of Tsukuba
Watermelon	CRISPR-Cas9	Agronomic	Disease resistance	China, USA	National Watermelon and Melon Improvement Center, Beijing Academy of Agricultural and forestry Sciences, Key Laboratory of Biology and Genetic Improvement of Horticultural Crops; Beijing Key Laboratory of Vegetable Germplasm Improvement, Beijing; Zhengzhou Fruit Research Institute, Chinese Academy of Agricultural Sciences; Boyce Thompson Institute for Plant Research, Cornell U; Lingnan Guangdong Laboratory of Modern Ag, Genome Analysis Laboratory of the Ministry of Ag, Agricultural Genomics Institute at Shenzhen, Chinese Academy of Agricultural Sciences; US Dept. of Ag, Ag Research Service, Plant Genetic Resources Conservation Unit, Georgia; USDA—Ag Research Service, U.S. Vegetable Laboratory, S. Carolina; Key Laboratory of Biology and Genetic Improvement of Horticultural Crops of the Ministry of Ag, Sino-Dutch Joint Laboratory of Horticultural Genomics, Institute of Vegetables and Flowers, Chinses Academy of Ag Sciences, Beijing; USDA Ag Research Service, Robert W. Holley Center, New York
Wild tomato	CRISPR-Cas9	Food & feed	High antioxidant content	Brazil, Germany, USA	Several universities from countries of origin

Adapted from: FAO (2022), Kumar et al. (2022)

References

Alvarez, J. (2021, April 15). *New CRISPR technology offers unrivaled control of epigenetic inheritance.* Research, University of California San Francisco. Retrieved January 7, 2024, from https://www.ucsf.edu/news/2021/04/420306/new-crispr-technology-offers-unrivaled-control-of-epigenetic-inheritance

Baltes, N. J., & Voytas, D. F. (2014). Enabling plant synthetic biology through genome engineering. *Trends in Biotechnology, 33*, 120–131.

Boch, J., Scholze, H., Schornack, S., Landgraf, A., Hahn, S., Kay, S., Lahaye, T., Nickstadt, A., & Bonas, U. (2009). Breaking the code of DNA binding specificity of TAL-type III effectors. *Science, 326*, 1509–1512.

Doudna, J. A., & Charpentier, E. (2014). Genome editing. The new frontier of genome engineering with CRISPR-Cas9. *Science, 346*(6213), 1258096.

Entine, J., Felipe, M. S. S., Groenewald, J. H., et al. (2021). Regulatory approaches for genome edited agricultural plants in select countries and jurisdictions around the world. *Transgenic Research, 30*, 551–584. https://doi.org/10.1007/s11248-021-00257-8

FAO (Food and Agriculture Organisation). (2022). *Gene editing and agrifood systems.* FAO (Food and Agriculture Organisation). Retrieved January 3, 2024, from https://doi.org/10.4060/cc3579en

Friedrichs, S., Takasu, Y., Kearns, P., Dagallier, B., Oshima, R., Schofield, J., & Moreddu, C. (2019). An overview of regulatory approaches to genome editing in agriculture (conference report). *Biotechnology Research and Innovation, 3*, 208–220.

Gao, C. (2018). The future of CRISPR technologies in agriculture. *Nature Reviews Molecular Cell Biology, 19*, 275–276.

Gatica-Arias, A. (2020). The regulatory current status of plant breeding technologies in some Latin American and Caribbean countries. *Plant, Cell, Tissue and Organ Culture, 141*, 229–242.

International Service for the Acquisition of Agri-biotech Applications (ISAAA). (2023). *Pocket K No. 59: Plant breeding innovation: TALENs.* Retrieved December 15, 2023, from https://www.isaaa.org/resources/publications/pocketk/59/default.asp

Karvelis, T., Gasiunas, G., Young, J., Bigelyte, G., Silanskas, A., Cigan, M., & Siksnys, V. (2015). Rapid characterization of CRISPR-Cas9 protospacer adjacent motif sequence elements. *Genome Biology, 16*, 253. https://doi.org/10.1186/s13059-015-0818-7

Kumar, D., Yadav, A., Ahmad, R., Dwivedi, U. N., & Yadav, K. (2022). CRISPR-based genome editing for nutrient enrichment in crops: A promising approach toward global food security. *Plant Genomics, 13*. https://doi.org/10.3389/fgene.2022.932859

Laaninen, T. (2020). *New plant-breeding techniques: Applicability of EU GMO rules.* European Parliamentary Research Service. Retrieved October 13, 2023, from https://www.europarl.europa.eu/RegData/etudes/BRIE/2020/659343/EPRS_BRI(2020)659343_EN.pdf

Lassoued, R., Hesseln, H., Phillips, P. W. B., & Smyth, S. J. (2018). Top plant breeding techniques for improving food security: An expert Delphi survey of the opportunities and challenges. *International Journal of Agricultural Resources, Governance and Ecology, 14*(4), 321–337.

Lassoued, R., Phillips, P. W. B., Macall, D. M., Hesseln, H., & Smyth, S. J. (2021). Expert opinions on the regulation of plant genome editing. *Plant Biotechnology Journal, 14*(4), 321–337. https://doi.org/10.1111/pbi.13597

Liang, Z., Zhang, K., Chen, K., & Gao, C. (2014). Targeted mutagenesis in Zea mays using TALENs and the CRISPR/Cas system. *Journal of Genetics and Genomics, 41*, 63–68.

Lyzenga, W. J., et al. (2021). Advanced domestication: harnessing the precision of gene editing in crop breeding. *Plant Biotechnology Journal, 19*, 660–670.

Moscou, M. J., & Bogdanove, A. J. (2009). A simple cipher governs DNA recognition by TAL effectors. *Science, 326*, 1501–1501. https://doi.org/10.1126/science.1178817

NASEM (National Academies of Science Engineering and Medicine). (2016). *Genetically engineered crops: Experiences and prospects.* National Academies Press. https://www.ncbi.nlm.nih.gov/books/NBK424553/

NASEM (National Academies of Science Engineering and Medicine). (2017a). *Preparing for future products of biotechnology*. NASEM (National Academies of Science Engineering and Medicine). https://doi.org/10.17226/24605, https://www.ncbi.nlm.nih.gov/books/NBK442207/

NASEM (National Academies of Science Engineering and Medicine). (2017b). *Human genome editing: Science, ethics, and governance*. E Glossary. National Academies Press. https://www.ncbi.nlm.nih.gov/books/NBK447273/

NASEM (National Academies of Science Engineering and Medicine). (2020). *Next steps for functional genomics: Proceedings of a workshop*. National Academies Press. https://nap.nationalacademies.org/read/25780/chapter/1

Novak, S. (2019). Plant biotechnology applications of zinc finger technology. In S. Kumar, P. Barone, & M. Smith (Eds.), *Transgenic plants. Methods in molecular biology* (Vol. 1864, pp. 295–310). Humana Press. https://doi.org/10.1007/978-1-4939-8778-8_20

Nunez, J. K., et al. (2021). Genome-wide programmable transcriptional memory by CRISPR-based epigenome editing. *Cell, 184*(9), 2503–2519.

Public research and regulation initiative (PRRI). (2023). *Site directed nuclease (SDN) genome editing*. Retrieved on October 19, 2023 from https://prri.net/scientific-topics/new-breeding-techniques/genome-editing/site-directed-nuclease-sdn-genome-editing

Sander, J. D., & Joung, J. K. (2014). CRISPR-Cas systems for editing, regulating and targeting genomes. *Nature Biotechnology, 32*, 347–355.

Seyran, E., & Craig, W. (2018). New Breeding Techniques and their possible regulation. *AgBioforum, 21*(1), 1–12.

Silva, G., Poirot, L., Galetto, R., Smith, J., Montoya, G., Duchateau, P., & Paques, F. (2011). Meganucleases and other tools for targeted genome engineering: Perspectives and challenges for gene therapy. *Current Gene Therapy, 11*, 11–27.

Tröder, S., & Zevnik, B. (2022). History of genome editing: From meganucleases to CRISPR. *Laboratory Animals, 56*(1), 60–68.

Williams, C. (2021, June 3). Gene editing and agriculture. *Farming Connect*. Retrieved on June 23, 2023, from https://businesswales.gov.wales/farmingconnect/news-and-events/technical-articles/gene-editing-and-agriculture

Chapter 3
How Are Gene Editing Technologies Regulated in the Agrifood System?

In the end, every regulatory system is precautionary, because you don't give automatic approvals.—Informant 6 (NGO representative)

3.1 Introduction

The central question that guides this book is why has gene editing—and specifically with the advent of the CRISPR-Cas9 system—not become the big breakthrough in agricultural biotechnology it was touted as when discovered in 2012? Despite its less-than-ideal introduction into the world, CRISPR-Cas9 has made many inroads in agricultural research and the development of marketable agrifoods. However, there is currently no international consensus on how gene edited organisms and products should be regulated. Countries have the autonomy to determine if gene edited agrifoods fall under the regulatory framework of the Cartegena Protocol on Biodiversity (CPB) and the Convention on Biological Diversity (CBD) and whether or not they are included in the definition of a Living Modified Organism (LMO).

In this chapter, we examine the role of regulation in the management and use of gene editing in the agrifood system. Many regulatory typologies have developed over the years, placing countries into specific groups based on their adherence to the Cartagena Protocol and its definition of GMOs and LMOs. Documentation of the evolution of regulations covering gene editing are widely available and referenced here (some examples include Laaninen, 2020; Shukla-Jones et al., 2018; NASEM, 2017; Ahmad et al., 2021; FAO, 2022a, b; Vora et al., 2023). Generally speaking, regulatory systems fall into five categories: light, strong/prohibited, proposed, modified and no regulation. The embeddedness of the Cartagena Protocol in regulatory frameworks largely determines where gene editing falls in terms of its allowable uses in the agrifood sector (research & development, environmental release, commercialization).

This chapter synthesizes the general issues which countries that currently regulate gene editing are addressing, rather than providing detailed analyses of the responsibility and approach of each national regulatory agency, which is beyond the scope of the chapter. Details regarding the technical responsibilities of national

L. F. Clark, J. E. Hobbs, *International Regulation of Gene Editing Technologies in Crops*, SpringerBriefs in Environmental Science, https://doi.org/10.1007/978-3-031-63917-3_3

regulatory agencies for regulating gene edited agrifoods can be found in various well-researched publications and websites such as the Genetic Literacy Project's 'Global Gene Editing Regulation Tracker' and are not repeated here (see Entine et al., 2021; Ahmad et al., 2021; Vora et al., 2023).[1] Instead, this chapter provides a snapshot of the current country-specific approaches to regulating gene edited crops and agrifoods to provide a better understanding of how regulatory frameworks may evaluate the biosafety of new plant breeding techniques on the horizon. Since many countries use their biosafety protocols for biotechnology as the foundation for regulating gene editing, this chapter assesses the state of regulation for agricultural biotechnology as it pertains to gene editing.

Over the last decade, there have been ongoing international discussions seeking legal clarity pertaining to the status of genome editing and derived products. Key regions around the globe have adopted different positions (Lassoued et al., 2021). Harmonized regulatory blocs have emerged in South America, Central America, Africa, North America, and Europe, though within regions there are multiple approaches to regulating gene edited agrifoods (see Sect. 3.3). These are functional agreements that allow for countries in the same geographic region to harmonize aspects of their regulatory frameworks to facilitate trade. We begin by discussing the current state of regulatory frameworks, then how each regulatory regime assesses environmental, human and animal health, and biosafety risks.

3.2 International Organizations and the Regulation of Gene Edited Organisms

The Organisation for Economic Co-operation and Development (OECD) Conference on 'Genome Editing: Applications in Agriculture – Implications for Health, Environment and Regulation' took place in Paris in 2018. It explored the "safety and regulatory considerations raised by genome edited products, with the aim to favour a coherent policy approach to facilitate innovation involving genome editing and will bring together policy makers, academia, innovators and other stakeholders involved in the topic."[2] It brought together relevant stakeholders (scientific experts, regulators, company representatives, etc.) from over 35 countries (OECD, 2018; Friedrichs et al., 2019). Participants in the Conference expressed the importance of fashioning regulatory approaches for genome editing to achieve policy objectives that consider precaution and innovation through better communication (Menz et al., 2020: 13). Though some of the countries discussed here enacted legislation prior to the OECD meeting, providing an overview of the global governance structures

---

[1] The Genetic Literacy Project (2023) 'Global Gene Editing Regulation Tracker' https://crispr-gene-editing-regs-tracker.geneticliteracyproject.org/

[2] See OECD website for full details. https://www.oecd.org/environment/genome-editing-agriculture/

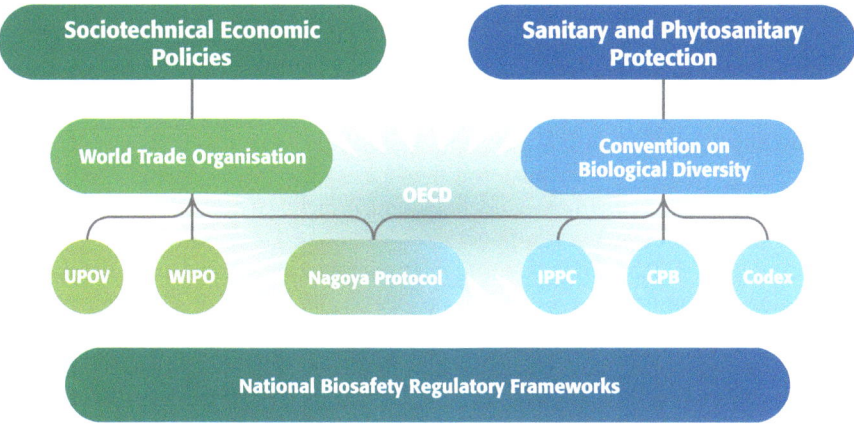

SOURCE: AUTHORS' COMPILATION; FAO, 2022: 30

Fig. 3.1 Governance of gene edited foods. (Adapted from FAO, 2022a)

pertaining to gene edited agrifoods helps to add context to the brief discussions below. Regulatory frameworks address scientific and sanitary considerations related to gene edited products. Though the Cartagena Protocol defines GMOs as LMOs, policy interpretations of what constitutes a GMO vary widely.[3] The arrival of gene editing in the biotechnology sphere has not necessarily resulted in opening a significant new approach to regulation, but rather entails addressing biosafety concerns within existing frameworks. In this section we discuss the roles of various international organizations in the governance of gene edited agrifoods.

Figure 3.1 shows the major actors in the global governance space and demonstrates how these actors interact with one another. The organizations' mandates fall under two main areas: those covering sanitary and phytosanitary protections,[4] and those covering sociotechnical and economic policies.[5] However, the interactions of organizations are not limited to those in the same category. All the organizations in Fig. 3.1 have some role in setting standards, developing consensus-based protocols, and facilitating global dialogue on the scientific, economic, and social implications

---

[3] According to the Convention on Biological Diversity, A Living Modified Organism (LMO) is defined in the Cartagena Protocol on Biosafety as "any living organism that possesses a novel combination of genetic material obtained through the use of modern biotechnology. The Protocol also defines the terms 'living organism' and 'modern biotechnology' (see Article 3). In everyday usage LMOs are usually considered to be the same as GMOs (Genetically Modified Organisms), but definitions and interpretations of the term GMO vary widely." For more information see https://bch.cbd.int/protocol/cpb_faq.shtml

[4] This includes the Convention on Biological Diversity, International Plant Protection Convention, World Organisation for Animal Health, and the Codex Alimentarius.

[5] This category includes the World Trade Organisation, The Organisation for Economic Cooperation and Development, Nagoya Protocol, World Intellectual Property Organisation, International Union for the Protection of New Varieties of Plants.

of innovative technologies in the agrifood system. Countries are signatories to certain protocols, or members of organizations, such as the World Trade Organisation (WTO) and/or OECD. In terms of decision-making, the governance space is horizontally oriented, with multi-directional interactions between and amongst organizations. There are several international organizations that guide these decisions, depending on whether the country is a member/signatory and therefore subject to the organization's regulations, guidelines and principles: the OECD, the World Trade Organisation (WTO), the Food and Agriculture Organisation/World Health Organisation's (FAO/WHO) Codex Alimentarius standards and codes, the International Plant Protection Convention (IPPC), World Organisation for Animal Health (WOAH), the Cartagena Protocol on Biosafety (CPB).

The OECD is at the forefront of the global dialogue regarding gene edited agrifoods. It gathers stakeholders to assess change, and potential responses to changes. The OECD is the central hub in the governance network, that other organizations look to for policy guidance, data on policy development and socio-economic considerations related to gene edited agrifoods. The OECD works towards consensus-based policy considerations and standards setting for gene edited agrifoods. The OECD acts as an information clearing house and knowledge hub and is a valuable source of information for regulators. It has an ongoing working party made up of country representatives that has been discussing issues regarding harmonization in biotechnology, safety of novel food and feeds since 1982. The OECD working party guideline documents provide detailed content on issues covered by the other organizations, as discussed below (Friedrichs et al., 2019).

The most important international guidelines relating to the global governance of gene edited agrifoods are the Convention on Biological Diversity (CBD)[6] and the Cartagena Protocol on Biosafety (CPB). The Cartagena Protocol on Biosafety to the Convention on Biological Diversity (a supplement to the CBD) was agreed upon in 2000, coming into force in 2003. It covers guidelines for risk assessments, environmental/biodiversity, and human health. Currently, the CPB has 173 signatories, most of which base the language used in their national regulations on the language in the Protocol. Its guidelines govern the transboundary movement of LMOs. The Nagoya Protocol (a supplementary agreement to the 1992 Convention on Biological Diversity) coined the definition for 'Living Modified Organisms' and 'food derived from modern biotechnology'. The Nagoya Protocol is a legal framework that lays out predictable international access to local genome sources. These definitions are used to establish if the treaty and related domestic regulation apply to gene edited organisms.

The FAO/WHO have also established guidelines for gene edited agrifoods. Joint FAO/WHO Food Standards Programs such as the Codex Alimentarius Guidelines (Codex), define 'modern biotechnology' based on the CPB definition, but have not adopted the CPB's definition of LMOs that includes GMOs. The guidelines include

---

[6] The Convention on Biological Diversity (CBD) was finalized in1992, entering into force the following year (CBD, 2018).

references to food safety assessments, which include principles for risk assessment, for conduct of food safety, plants, animals, and microorganisms on labelling. The guidelines also include standards on analytic methods to evaluate products of biotechnology, and how to assess the equivalence of sanitary measures associated with food inspection and certification systems.

The International Plant Protection Convention (IPPC), an intergovernmental treaty established in 1951 and overseen by the United Nations, has been ratified by 184 countries. It includes standards and procedures for identifying pests that threaten plant health, assessing risk, and determining the strength of regulatory measures to limit unintended spread of pests, invasive species, etc. Most countries have guidelines to assess risks of pests that threaten plant health.

In terms of differences in national regulations that impede international trade, the SPS (Sanitary-Phytosanitary Agreement) of the WTO includes agreement on the use of scientific principles as the basis of SPS measures that restrict trade for the protection of plant, human, and animal health. The WTO Agreement on Trade-Related Aspects of Intellectual Property Rights (TRIPS) sets out standards for the protection of intellectual property, including patents, trademarks, geographical indications, and copyright (WTO, 1994). The World Intellectual Property Organization (WIPO) administers treaties on intellectual property, which is a prime concern in the increasingly complex mix of biotechnology techniques and applications. But ultimately, each national framework for gene edited agrifoods is autonomous. Signatory countries are obliged to abide by the protocols and agreements they sign, but how standards and principles are enforced is largely up to the state-level governing bodies.[7]

Soon after the discovery of CRISPR-Cas9 in 2012, there were many questions raised about this novel gene editing technique and about CRISPR's impact on plant breeding, and national security (e.g., CRISPR's potential use in biological warfare), and how to regulate it. The increasing experimentation with CRISPR-Cas9 in agrifood raised questions in many jurisdictions about how to regulate this novel technique, necessitating the OECD meeting in 2018. Policy makers were trying to build some consensus on whether to use their respective countries' pre-existing national regulatory frameworks covering GMOs, or to develop a different policy framework to accommodate agricultural biotechnology techniques that do not use foreign DNA (SDN1 and SDN2) and are not considered transgenic. Regulators were then faced with, what Asquer and Krachkovkaya (2021: 1114) refer to, as 'response uncertainty'. Response uncertainty, "…arises from the inability to predict, which regulatory tools are more suitable to the specific features of the emerging technology." Response uncertainties include considerations over whether restrictive regulations limit R&D, innovation, and domestic business or whether permissive regulations could result in unintended consequences, risks to health and safety and/or harms to humans, animals and/or the environment (Asquer & Krachkovskaya, 2021). As

---

[7] Signatories are obliged to enact and enforce articles in the Agreements or be subject to WTO dispute settlement mechanisms.

discussed in the next section, there have been several ways regulatory frameworks have responded to the response uncertainty presented by CRISPR-Cas9.

3.3   State-Level Regulatory frameworks for Gene Edited Foods

Countries have taken different paths towards regulating gene edited agrifoods. Figuring out the right way to combine the benefits of innovation, economic development, health and safety, and social license is a difficult task, especially when the functions of nucleases that can edit the genome to achieve desirable traits are discovered every year. Some countries have been influenced by the positions taken by neighbouring countries in their geographic regions and by their trading partners. Others are in the midst of crafting their regulatory frameworks, seeking guidance from regulators in other jurisdictions who have experience with developing standards and protocols. A third group of countries are currently developing regulatory frameworks for all forms of biotechnology in agriculture, covering GMOs and gene edited agrifoods where none existed before, as is the case for several developing countries across Africa. There is also a host of countries that have no publicized regulations, and no active policy discussions.

Here, we briefly discuss the five general categories of regulatory protocols for gene edited agrifoods that countries fall into: lightly regulated, strongly regulated/prohibited, proposed regulation, modified process, and no regulations/no active policy discussions. Table 3.1 outlines in detail where 53 countries fall in terms of whether they are signatories of the CPB, the date of the first regulatory decision regarding gene editing, and the type of regulation. Table 3.1 also categorizes each country into whether it follows a 'product' or 'process' based regulatory approach. The product-based approach is premised on the idea that the 'risks posed by agrifoods are a function of biological characteristics and the specific genes that have been used,' while the process-based approach to evaluating the risks posed by products of biotechnology are based on 'processes employed in their development' (Tagliabue, 2017: 3; Prakash et al., 2014). Information is not readily available for all countries, nevertheless, the Table provides a reference point for the discussions in this chapter. Table 3.1 summarizes the regulatory landscape for gene edited crops across countries as of early 2024.

Generally speaking, Site Directed Nucleases interventions, where foreign DNA is inserted into a host genome (SDN3), are considered GMO (transgenics) (Sprink et al., 2016). These are considered 'novel combinations of genetic material', the language used in the CPB's definition of LMO. This wording is used in some national regulations to define GMOs. As discussed in Chap. 2 (Sect. 2.2.1), SDN1 does not use foreign genes to create edits. It harnesses the natural repair mechanisms that exist in the organism's cell. It has similarities to the general mechanisms and results of breeding techniques that involve chemical or radiological

Table 3.1 The regulatory landscape for gene edited crops

Country/ Region	Cartagena protocol on biosafety	Date of first regulatory decision	Type of regulation[a]	Product or process evaluation	Notes
Argentina	No	2019	Light regulation	Product	Gene edited crops are deregulated if they have no transgene; case by case assessment
Australia	No	2019	Light regulation	Product	Gene edited crops allowed when NHEJ machinery repair the break naturally; edited crops are regulated if donor template or foreign genetic material inserted using editing tools
Bangladesh	Ratified	2012	Light regulation	N/A	If no foreign DNA is present, gene edited plants are evaluated for safety the same as conventionally produced crops (new decision, March 2024)
Belize	Accepted	N/A	No regulations/ No active policy discussions	N/A	Moratorium that prevents environmental release of GMOs
Bolivia	Ratified	2019	Strongly regulated/ Prohibited	Process	Government approved 2 gene editing events for soybeans in 2019 and considering cotton and corn
Brazil	Accepted	2018	Light regulation	Product	Gene edited crops are deregulated if they have no transgene; case by case assessment
Burkina Faso	Ratified	N/A	Proposed regulation	N/A	
Canada	No	2022	Light regulation	Product	Gene edited crops are considered a 'fast version' of traditional breeding, case by case assessment
Chile	No	2018	Light regulation	Product	Gene edited crops are deregulated if they have no transgene; case by case assessment
China	Approved	2022	Modified process	Process	Ongoing research; regulations in development
Columbia	Ratified	2018	Light regulation	Product	

(continued)

Table 3.1 (continued)

Country/ Region	Cartagena protocol on biosafety	Date of first regulatory decision	Type of regulation[a]	Product or process evaluation	Notes
Costa Rica	Ratified	N/A	Proposed regulation	Product	November 2023 update to biosafety regulations reduces barriers to common applications of modern biotech
Ecuador	Ratified	2019	Light regulation	Product	
Egypt	Ratified	N/A	No regulations	N/A	
El Salvador	Ratified	N/A	Light regulation	Product	
Ethiopia	Ratified	N/A	Proposed regulation	N/A	
European Union[b]	Approved	2018	Strongly regulated/ prohibited	Process	Dir. 18/2001/EC (2001) after court decision in case C-528/16; EC has initiated a process to develop a regulatory proposal for plants resulting from the application of targeted mutagenisis and cisgenesis
Ghana	Accepted	N/A	Proposed regulation	N/A	
Guatemala	Accepted	2019	Light regulation	product	
Honduras	Ratified	2019	Light regulation	product	
India	Ratified	2020	Light regulation	product	Has defined policy framework to regulate GM crops, no guideline for the release of edited crops yet
Indonesia	Ratified		Proposed regulation	N/A	
Israel	No		Light regulation	Product	
Japan	Accepted	2019	Light regulation	Product	Handling of organisms obtained via genome editing tech that do not fall under GMO category as defined by Cartagena Act (2019)
Kenya	Ratified	TBD	Light regulation	Product	
Malawi	Ratified	N/A	Light regulation	Product	

(continued)

Table 3.1 (continued)

Country/ Region	Cartagena protocol on biosafety	Date of first regulatory decision	Type of regulation[a]	Product or process evaluation	Notes
Mexico	Ratified	N/A	Limited research, no regs	N/A	Largely restricted, no active discussion
Mozambique	Ratified		Proposed regulation	N/A	
Myanmar	Ratified	N/A	No regulations/ No active policy discussions	N/A	Recognizes ASEAN Guidelines on Risk Assessment of Agriculture-Related GMOs
New Zealand	Ratified	2019	Strongly regulated/ prohibited	Process	Gene editing deregulated if no new genetic material is added
Nigeria	Ratified	TBD	Light regulation	Product	
Northern Ireland	Ratified				Complies with EU Legislation under the current terms of the Protocol on Ireland/Northern Ireland (covered in Annex II)
Norway	Ratified	2018	Strongly regulated/ prohibited	N/A	No active discussions
Pakistan	Ratified		No regulations/ No active policy discussions	N/A	Research allowed, but no trials or gene edited crop has progressed through the regulatory system
Paraguay	Ratified	2019	Light regulation	Product	
Peru	Ratified	N/A	Strongly regulated/ prohibited	Process	
Philippines	Ratified		Light regulation	Product	
Republic of Korea	Ratified	TBD	Proposed regulation	Product	
Russia	No	2020	Proposed regulation	Product	Gene editing for research purposes only; government investing in gene-editing programs aiming to develop 10 new varieties of crops

(continued)

Table 3.1 (continued)

Country/Region	Cartagena protocol on biosafety	Date of first regulatory decision	Type of regulation[a]	Product or process evaluation	Notes
Singapore	No		Proposed regulation		
South Africa	Accepted	N/A	Strongly regulated/prohibited	Process	Ongoing appeal procedure to modify regulations
Sudan	Accepted	N/A	No regulations/No active policy discussions	N/A	
Switzerland	Ratified	N/A	Proposed regulation	N/A	Ongoing research; Regulations in development
Taiwan	No	N/A	Proposed regulation	N/A	
Turkey	Ratified		Strongly regulated/prohibited	Process	
Uganda	Ratified	N/A	Proposed regulation	N/A	
Ukraine	Accepted	N/A	Proposed regulation	N/A	In the process of developing regulatory framework
United Kingdom	Ratified	see notes	Proposed regulation	Product	Genetic Technology (Precision Breeding) bill was announced excluding genetic technologies (May 2022). Only applies to England (Scotland and Wales have indicated they do not wish to adopt the measure to change the definition of GMO and the intro of a new framework for food & feed)
United States	No	2015	No unique regulation	Product	
Uruguay	Ratified	2018	Proposed: no unique regulation	Product	Gene editing used in research
Venezuela	Ratified		Strongly regulated/prohibited	Process	2015 Seeds Act bans cultivation of GMO crops
Vietnam	Accepted	TBD	Proposed regulation	N/A	

(continued)

Table 3.1 (continued)

Country/ Region	Cartagena protocol on biosafety	Date of first regulatory decision	Type of regulation[a]	Product or process evaluation	Notes
Zambia	Accepted		Proposed regulation		
Zimbabwe	Ratified		Proposed regulation		

Sources: Convention on Biological Diversity (2023), Genetic Literacy Project (2023), FAO (2022b), USDA (2019), South Asia Biosafety Program (2021), Gatica-Arias (2020), Entine et al., (2021), Jones et al. (2022), Tiwari (2021), Menz et al. (2020), Ahmad et al. (2021)
Notes:
For up-to-date country-level regulatory changes, see Convention on Biological Diversity (https:// bch.cbd.int/protocol/parties/); The Genetic Literacy Project (https://crispr-gene-editing-regs-tracker.geneticliteracyproject.org/)
[a]See Fig. 3.2 for more information on Type of Regulations
[b]European Union includes Austria, Belgium, Bulgaria, Croatia, Cyprus, Czech Republic, Denmark, Estonia, Finland, France, Germany, Greece, Hungary, Ireland, Italy, Latvia, Lithuania, Luxembourg, Malta, Netherlands, Poland, Portugal, Romania, Slovakia, Slovenia, Spain, Sweden

mutagenesis. Because of different classifications of techniques, there are different views on how each technique should be appropriately regulated. There are different opinions on how each SDN technique should be defined in regulations. Some argue that SDN1 does not meet the legal definition of LMO/GMO because there are no foreign genes inserted into the genome, such as in the case of Canada. Others, such as the European Union, consider SDN1 techniques as products of 'modern biotechnology' and therefore having the same risk profile as GMOs (and are regulated as such). Some genome editing techniques were not discovered until after the establishment of biosafety regulatory frameworks in the 2000s. An update to regulation is then required to include the products that did not previously exist (Eriksson et al., 2019). Some have argued that it makes more scientific sense to move away from product vs. process-based definitions and evaluate the safety of the gene edited organism or products based on traits, creating a trait-based approach (Qaim, 2009; Gould et al., 2022). This approach has not yet been embraced by any one country.

Asquer and Krachkovskaya (2021: 1122) see global governance of CRISPR gene editing technologies as falling into two main camps: The EU approach and the US approach, with nuanced variations in the middle. They argue that the primary difference in how the EU and the US agencies responded to the advent of CRISPR as the dominant gene editing technique is that the EU interpreted the response uncertainty about CRISPR-Cas9 by subjecting the novel technique to existing regulatory frameworks rather than undertaking any institutional review and adjustments of regulatory practices. The EU is in the process of reviewing current Directives covering biotechnology as they apply to gene editing. The US chose the latter pathway (institutional review and adjustments of regulatory practices), as did other countries such as Canada and Australia.

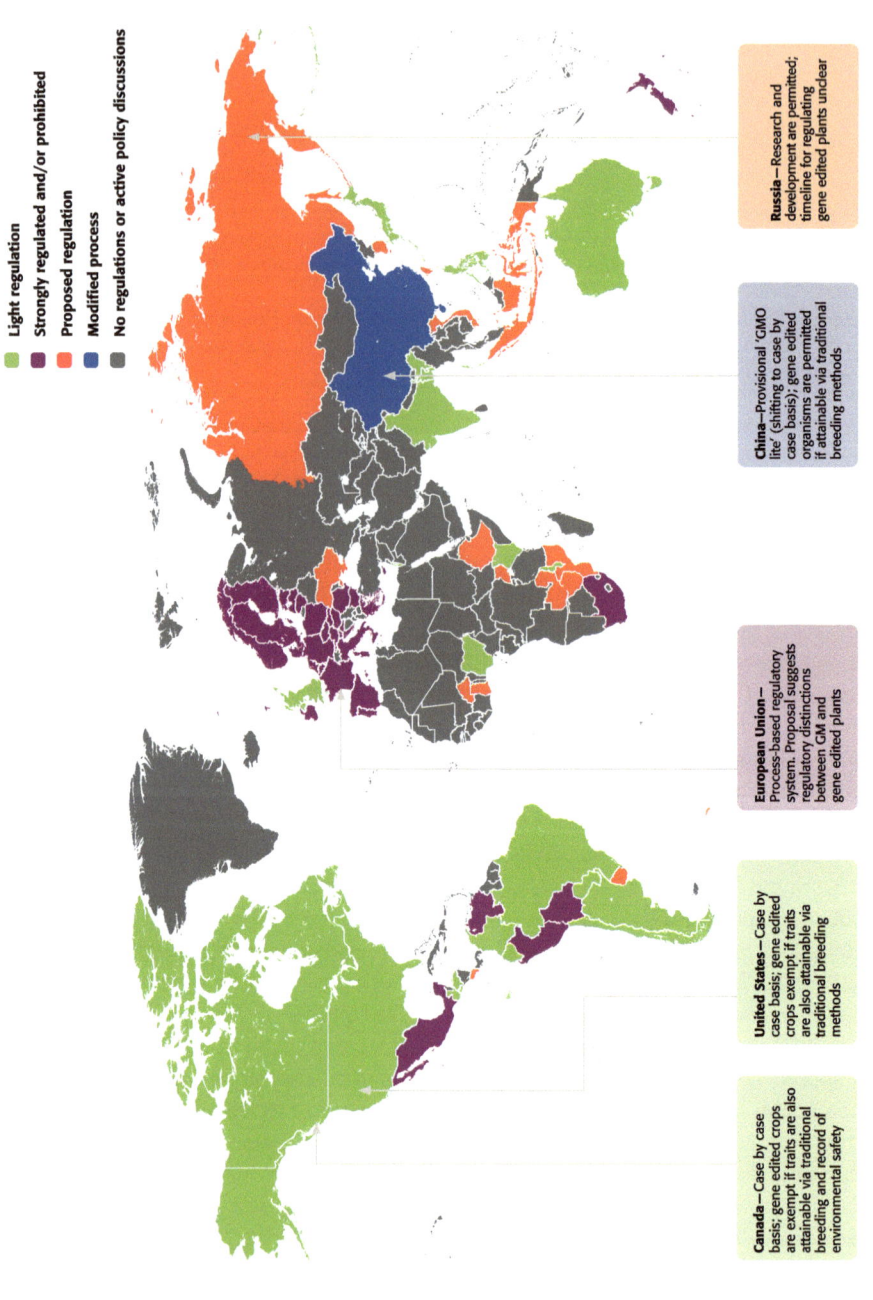

Fig. 3.2 Current regulatory environment for gene edited plants. (Adapted from Chou, 2023; Lawrence, 2023)

Experts have developed many typologies to categorize regulatory frameworks for gene edited agrifoods (see Entine et al., 2021; Chou, 2023; Lawrence, 2023; Ahmad et al., 2021). But they all refer to five distinct approaches to regulating gene editing agrifoods: light regulation, strong regulation, proposed regulation, a modified process, and no regulation/policy discussions (see Table 3.1, column 4). Some countries are in the proposed regulation stage of development, and currently do not naturally fit into any of the above categories. As such, we have categorized them under separate headings (proposed regulations, no regulations/no active policy discussions).

Other systems of classification divide regulatory systems into process vs product-based forms of risk assessment (see Table 3.1, column 5). A product-based approach means that regulations are based on risk assessment of the product, regardless of the process by which products or organisms were developed, while in a process-based approach a different set of regulations applies depending on the process used (e.g. gene editing). According to Friedrichs (2019: 209), there are several jurisdictions using the 'process-based' model that are currently reviewing the scope of their definitions of whether gene edited agrifoods are regulated as GMOs or LMOs, including New Zealand, and the European Union. Product triggered regulatory systems, most notably Argentina, Brazil, Canada, and the United States consider the novelty of the trait on a case-by-case basis no matter what process was used to achieve the novel trait.

Figure 3.2 summarizes the current global regulatory environment for gene editing into five categories: light regulation, strongly regulated and/or prohibited, proposed regulation, modified process, and no regulations/active policy discussions.[8]

3.3.1 Light Regulation

The light regulation category includes countries that have defined GMO exemptions for gene edited agrifoods (See Fig. 3.2, green shading). The regulatory system differentiates between gene edited agrifoods and GMOs. Australia and New Zealand have harmonized their regulatory frameworks under the GMO regulation law. Food Standards Australia New Zealand (FSANZ) is an independent statutory authority that enforces the Australia New Zealand Food Standards Code ('the Code'), which is a collection of enforceable food standards. They both use process as a trigger, however organisms using SDN1 are not considered GMOs. Australia amended its Gene Technology Regulations (2001) in 2019 which states that it will not regulate the use of gene editing techniques that do not introduce new genetic material

---

[8] Changes to regulations covering gene editing crops in smaller countries with smaller agriculture sectors are occurring rapidly as countries such as Canada, the US and Australia with much larger agriculture sectors have decided to regulate gene edited crops differently than GMOs. Thus, Fig. 3.2 may not reflect the most up-to-date categories for certain countries, especially those in the process of developing regulations.

(SDN1). SDN1 plants are no longer considered GMOs and are no longer regulated under the Gene Technology Act 2000 and are now regulated by the Department of Agriculture, Water and the Environment. Despite the shared governing responsibilities under the FSANZ of GMOs standards, however, New Zealand continues to evaluate the biosafety of gene edited plants the same as GMOs (Ahmad et al., 2021).

Japan applies a case-by-case methodology to any organism with novel traits applying for regulatory approval. The Ministry of Health, Labour and Welfare published final guidelines in 2020 (Laaninen, 2020). The Ministry of Environment determines whether the product falls outside of the scope of LMO. Gene edited agrifoods are not subject to GMO biosafety protocols if no foreign DNA is present in the final product. Japan has approved two agrifood products for commercialization: a GABA enriched tomato and sea bream fish were approved in 2021, which are both available to Japanese consumers (Menz et al., 2020: 14).

Nigeria's Biosafety Management Agency has determined that if no foreign DNA is present in the organism, a non-GMO classification is given. Nigeria's regulatory framework shares similarities with Argentina, Australia, Brazil, Colombia, Israel, Japan, and Paraguay (Entine et al., 2021: 565). Kenya's National Biosafety Authority has indicated it will evaluate the biosafety of gene edited agrifoods on a case-by-case basis (Entine et al., 2021: 565). Final decisions will be based on the presence or absence of transgenes in the final product, which is similar to the framework introduced by Argentina in 2015 (Whelan & Lema, 2019).

Canada is also considered a 'light' regulator. It uses a product-based approach. Gene edited organisms are evaluated on a case-by-case basis but are not considered equivalent to GMOs with foreign DNA (Entine et al., 2021: 568). Section 3.4 provides a detailed discussion of the Canadian case. The United States, similar to Canada, takes a product-based approach to risk assessment, and finalized this into regulation in 2015. The United States Department of Agriculture (USDA), The Animal and Plant Health Inspection Service (APHIS), and the Food and Drug Agency (FDA) evaluate products for health and safety. Biosafety regulation is triggered by risk factors on a case-by-case basis. Regulatory frameworks are different for gene edited animals as opposed to plants. In 2020, a final rule updated and modernized the USDA biotechnology regulations under the Plant Protection Act. The 'Sustainable, Ecological, Consistent, Uniform, Responsible, Efficient' (SECURE) rule "brings USDA's plant biotechnology regulations into the twenty-first century by removing duplicative and antiquated processes in order to facilitate the development and availability of these technologies through a transparent, consistent, science based, and risk-proportionate regulatory system" (Laaninen, 2020: 8). The development of market-oriented traits, such as gene edited soybeans with higher levels of oleic acid, is led by the United States and China. The expectations the US has about future innovative technologies in agriculture largely informed how the regulatory system has responded to the growing popularity of CRISPR. This is what Asquer and Krachkovskaya (2021: 1122) refer to as the 'anticipatory response'.

Several South and Central American countries share similar regulatory profiles. Argentina has commercially grown GM crops since 1996 and uses the Cartagena Protocol's definition of LMO. The Ministry of Agriculture, Livestock and Fisheries evaluates applications on a case-by-case basis. Gene edited plants and animals are not considered GMOs because they do not have novel combinations of genetic material. After the Argentinian regulation was enacted, other Latin American countries developed their own set of regulations. Chile's Agriculture and Livestock Service aligned with Argentina and passed similar biosafety regulations to Argentina's in 2016. Paraguay (Ministry of Agriculture and Livestock, and Paraguay National Commission on Agriculture and Forestry Biosafety) shares some principles with Argentina. Both Colombia (Agriculture Institute) and Brazil enacted their legislation in 2018 (Whelan & Lema, 2019). The National Technology Biosafety Commission of Brazil shares similar approaches with Argentina and Chile. The regulatory framework goes further to include a list of techniques and genetic interventions that are not considered GMOs.

Paraguay, Ecuador, Honduras, and Guatemala passed their regulations in 2019. If foreign DNA is used in an organism, Ecuador's regulatory agency views this as a GMO, excluding SDN1 and cisgenesis (Entine et al., 2021: 552). Guatemala and Honduras (Guatemala Ministry of Agriculture, Livestock and Foodstuff, and Honduran National Service of Agrifood Health and Safety, respectively) have harmonized their regulations for GMOs. The regulations are based on a specific definition for 'novel combination of genetic material' and final product characteristics compared with conventional breeding products, essentially a 'case-by-case' basis for evaluation.

Israel's Plant Protection Services Administration takes a similar perspective on how gene edited plants are regulated. Plants without foreign DNA are classified as non-GMO as long as no foreign DNA is in the organism. Since leaving the EU, the United Kingdom (UK) has re-evaluated its regulatory stance on GMOs and is set to release updated guidelines. Part of the Genetic Technology (Precision Breeding) Act 2023, crops derived through precision breeding are no longer subject to the same regulations as GMOs in the UK. The Department of Environment, Food and Rural Affairs' definition of a GMO excludes organisms that have genetic changes that could occur naturally or achieved through traditional breeding.

As Informant 17, a research scientist working for an NGO, stated with respect to where India falls in the regulatory typology,

> where in most of the other countries, including India, 'process' is what is governing the GM definition.

There are many research programs and activities that are underway in India researching crop improvement using gene editing. India's Ministry of Environment, Forest and Climate Change, and the Ministry of Science and Technology regulates plants with novel traits derived from SDN1 or SDN2 as exempt from GMO regulations (which cover products of biotechnology that used SDN3). The ministries are currently examining whether all new technologies should be regulated as per existing regulatory frameworks. When new crop varieties come up for assessment using gene editing the regulatory framework will have to be firmly in place (Tiwari et al., 2021). As of 2023, regulatory amendments are still in process, though India's framework covering gene edited agrifoods closely adheres to the scope of other regulatory frameworks in the lightly regulated category.

Most recently, Bangladesh passed legislation for gene edited agrifoods in March 2024 that clearly places it in the lightly regulated category along with India. If applicants can demonstrate "… the absence of any exogenously introduced DNA and request confirmation through appropriate channels to register or release the plant following the same procedure as those used for conventionally bred counterparts" (Genetic Literacy Project, 2024a).

3.3.2   Strongly Regulated and Prohibited

This category includes countries that have regulatory systems that have no exceptions for gene edited agrifoods (see Fig. 3.2, purple shading). Gene edited foods are assessed for risk the same way as GMOs. There are several countries that have prohibited the environmental release and/or cultivation of GMOs and gene edited organisms. The EU via its EU Directive (Directive 01/18/EC) led by the European Food Safety Authority takes a precautionary approach. All organisms derived from mutagenesis are considered GMOs, however if mutagenesis is derived from conventional breeding techniques with a long history of safety it is exempt. The European Union is a global leader in gene editing research (but behind on market-oriented trait development), despite its strict policies (Modrzejewski et al., 2019).

Like the EU, Switzerland considers gene edited organisms as GMOs. In the EU, the advent of CRISPR-Cas9 in agrifood research triggered a regulatory response in the form of reaction by existing regulatory standards, what Asquer and Krachkovskaya (2021: 1123) refer to as a 'consequential response'. This consisted of a reactive approach to technologies such as gene editing. In July 2023, the EU Commission publicized a proposed amendment to the Directive covering all products of biotechnology. New Genomic Techniques (NGT) plants are split into two categories: naturally occurring or by conventional breeding; all other NGT plants are treated as GMOs that require risk assessments and authorization. If the current proposal is accepted, it would lift regulatory burdens on some gene edited agrifoods. A fast-track approval process is proposed that would apply to the second category of plants if they are more tolerant to climate change or require less water or fertilizer. Though Norway is not part of the European Union, it shares a highly restrictive position towards gene edited agrifoods. GMOs and gene edited agrifoods are restricted from environmental release, though scientists in Norway are pressing for modified restrictions for organisms derived using SDN1 techniques. According to the Genetic Literacy Project, "Norway law also demands the assessment of three non-safety categories: social benefit, sustainable and ethically sound products", going beyond its neighbouring EU regulations (Genetic Literacy Project, 2024b).

Mexico was on the pathway to establishing a product-based biosafety protocol, but reversed direction on gene edited agrifoods. Gene edited foods are evaluated based on the process and are considered equivalent to GMOs. Informant 10 (a research scientist) worked in Latin America on gene editing in agrifood and informed us that,

> when the [current Mexican] government came into power [in 2018] …they reject everything. They reject biotechnology, whether it is GM, whether it is genetic editing, genome editing, anything that has a hint of biotechnology and a lab they reject.

In addition to Mexico's current stance on gene edited agrifoods, Belize has a moratorium preventing gene edited organisms from release into the environment.

South Africa's Department of Agriculture, Land Reform and Rural Development use a risk assessment framework for GMOs that would also apply to new breeding techniques such as gene editing. It has opted for a tiered approach to assessments. South Africa's regulatory framework includes the EU-based definition for GMOs, which is somewhat divergent from how the Cartagena Protocol defines GMOs. Using phraseology such as "naturally occurring genetic variation" as a threshold to trigger biosafety protocols has garnered criticism, as has South Africa's chosen definition of GMO (Entine et al., 2021: 566).

Though part of the FSANZ agreement with Australia as discussed in the previous section, New Zealand continues to define gene edited organisms as GMOs. In 2014, there were discussions regarding changing how gene edited agrifoods were regulated but this was not successful. New Zealand continues to regulate gene edited agrifoods under the Hazardous Substances and New Organisms Act under the Environmental Protection Authority (Ahmad et al., 2021).

3.3.3 Proposed Regulation

The countries in this category as of early 2024 were in the midst of finalizing their regulations for gene edited agrifoods (Fig. 3.2, orange shading). There are ongoing policy discussions and draft proposals currently under review. This category includes Burkina Faso, Ethiopia, Uganda, Zimbabwe, Taiwan, Vietnam, South Korea, El Salvador, Thailand, Russia, and Uruguay (Chou, 2023). Some countries in this category are leaning towards regulating gene edited agrifoods the same way as GMOs, while others are seeking a simplified assessment procedure.

According to Informant 19 (a private sector representative), there are open discussions among several sub-Saharan African countries regarding harmonization of regulatory frameworks for gene editing agrifoods. They explain:

> The conversation is starting to happen on having a harmonized regulatory landscape. For example, Kenya, Uganda, Tanzania, South Sudan and a little bit of Rwanda. These conversations are happening since we trade closely with each other. Maybe it's best for us to come up with a more regional regulatory framework for these emerging technologies. But I would say for now, it's still fragmented. When you look at Kenya, for example, it could be a step ahead when it comes to genome editing and developing the regulatory framework for that innovation. And Uganda is still grappling with putting in place the regulatory framework for GMOs and so I would say countries are starting to have that conversation. But there's no

homogeneity in getting this to be operationalized. The concern is that it would help if regions had more harmonized regulatory frameworks. (Informant 19)

Costa Rica is also developing its own regulatory framework for gene edited agrifoods. The State Phytosanitary Service has procedures in place for determining whether a gene edited crop is considered a GMO. In November 2023, the country amended its biotechnology regulations to reduce barriers to common applications of modern biotechnology. It is expected that the first gene edited banana that is resistant to yield reducing fungal diseases will be on the market in Costa Rica by the end of 2024 (USDA, 2024). But as Informant 19 states, there are ongoing challenges for developing countries who are creating regulatory frameworks from scratch. They state:

> …developing countries…there's a challenge to have a regulatory landscape that is working for technologies that are pre-existing like genetic engineering. So before we even had closure on that, genome editing is coming in and it's creating what you might call 'a spanner in the works' where some policy makers are like, 'wait, wait, should we pause?' Formulating the regulatory landscape for GMOs and consider how we can incorporate these imagined technologies. So there's that challenge of whether to use existing regulatory frameworks for GMOs for genome editing or to completely start afresh with genome editing and because of that, it's affecting the speed at which the genome edited products and innovations may benefit the intended end users. So that's one of the key challenges I see, especially here in the developing world. (Informant 19)

The Russian government has invested substantially in gene editing research programs, according to Tiwari et al. (2021: 18) and initially planned to develop ten new varieties of gene edited crops. In 2019, a federal program was announced that included developing thirty new varieties of gene edited agrifoods by 2027 (Dobrovidova, 2019). The country has also prohibited the environmental release of GMOs and gene edited organisms; however, researchers have been allowed to experiment with these techniques for research purposes only as of 2016. Russia is currently running experiments using gene editing on barley, sugar beets, wheat, and potato. Russia is the largest producer of barley, and a leader in producing the other three crops. The announcement of funding for gene editing research into economically important crops for Russia raises some questions as to whether the regulatory framework will apply to these crops.

3.3.4 Modified Process

The most active country in genome editing research is China. However, it has not released any legal documents referencing its regulatory stance on gene editing organisms, but it is possible that China wants to have a product in hand before releasing regulations (Modrzejewski et al., 2019). China is also in the process of releasing its regulatory framework for gene edited agrifoods. However, its current legislation considers gene edited agrifoods GMOs, but with a simplified assessment procedure ('GMO light'). The Ministry of Agriculture and Rural Affairs allows gene edited plants that are determined to not present biosafety risks or risks to human health to be subject to a simplified registration procedure. It is predicted that China will lean towards the case-by-case basis for gene edited agrifoods risk and safety assessments, similar to Canada and the US, but not identical.

3.3.5   No Regulations

There are countries that do not have regulations for GMOs and there are no active policy discussions pertaining to how gene edited agrifoods should be regulated. There are also countries that currently do not have regulatory frameworks covering GMOs or gene edited agrifoods: Myanmar, Sudan, North Korea, Pakistan, and Egypt. This does not mean that gene edited crops can be released into the environment under any conditions. It simply means that these countries do not have regulatory frameworks covering gene edited plants yet and that there are no current, active discussions occurring to develop a framework. Countries that do not have active policy discussions are shaded grey in Fig. 3.2.

The influence of more established, larger trading partners is also something to consider, as it comes up in many discussions of regulatory frameworks emerging in the developing world. The precautionary position of Europe has exerted considerable influence on the regulatory frameworks of its smaller trading partners in Africa and elsewhere. Because many African countries export products to the EU, which strictly regulates GMOs and gene edited agrifoods, regulators in countries dependent on trade with the EU have had to consider the implications of taking a more open approach to gene editing, that resembles Canada, Australia, or Argentina's approaches. Informant 19, who engages with several African governments on genomic technologies in the agrifood system, has witnessed the strong influence the EU exerts on its African trading partners. They note:

> …you see that much of what is happening in Africa, in sub-Saharan Africa, for example is influenced by Europe's stand on the issue. If Europe is saying we are going to regulate edited products as GMOs that is going to have an effect on many countries in Africa and this is because of the complicated history of Europe and Africa…the biggest driver I see when it comes to the conversation on what regulatory frameworks should look like…Europe's approach to this is going to have a ripple effect in much of Africa.

As we can see, there is a diversity of approaches to regulating gene edited agrifoods. Some countries, namely in the same geographic region, have developed regulations that harmonize with their neighbors, as in South and North America. Divergent regulatory approaches result from different social, economic, and political realities. Different approaches to gene edited agrifoods as either considering them equivalent to GMOs or defined as LMOs complicates trade relationships in the agrifood system. A globally harmonized regulatory framework and guidelines is one suggestion put forth by experts in order to maximize the economic (and environmental) benefits of gene edited crops (Entine et al., 2021: 552).

3.4   Case Study: Canada

Canada's approach to regulating gene edited agrifoods has gained international praise for its focus on evaluating the end product rather than the techniques used in the plant breeding process. Therefore, it is worth delving a bit deeper into how the Canadian regulatory framework functions. For example, private sector representative Informant 8 stated that:

> Canada's regulatory approach, which is strongly supported by international science, going back, 20 years, focuses on the end result. When we're talking about plant breeding, it's the end result that matters.

Before we discuss how gene edited agricultural plants were declared to be 'non-novel' in the Canadian regulatory system in 2022, it is worth commenting on the basis for the Canadian gene editing policy position. The Government of Canada regulates plants, animal feed and human food separately under different sets of legislation. Therefore, the regulatory requirements are based on case-by-case evaluation of end products (the product-based approach). Table 3.2, modified from Friedrichs (2019: 212), shows how decision-making over particular agrifood products is divided up among the relevant agencies. The three agencies that are relevant to the discussion on gene edited agrifoods are the Canadian Food Inspection Agency (CFIA), Health Canada, and Agriculture and Agrifood Canada (AAFC). The table shows the regulatory agency, the regulations it enforces and what products are covered by the regulation.[9]

For this discussion, the roles of CFIA and Health Canada are of primary interest regarding gene edited agrifoods. In Canada, the decisions by Health Canada and CFIA regarding whether or not to treat an agrifood as novel is based on whether the product is considered a 'plant with novel traits' (PNT) in an agricultural context, or

Table 3.2 Authority, regulations covering novel agrifoods in Canada

Regulatory authority	Product	Regulation
Canadian Food Inspection Agency	Livestock feed	Feeds Act
	Seeds	Seeds Act
	Veterinary biologics	Health of Animals Act
Health Canada	Pesticides	Pest Control Act
	Novel foods, drugs, biologics, medical devices	Food and Drugs Act
Agriculture and Agrifood Canada	Non-regulatory considerations	Market access, industrial policy, socio-economic impacts, trade

Adapted from Friedrichs et al. (2019: 212)

---

[9] It is not an exhaustive table, but rather highlights the relevant areas for this discussion.

a novel food in the case of Health Canada.[10] The CFIA evaluates all novel plants using a product trigger regardless of the technology used to develop the trait, such as agricultural crops or feed for livestock. Health Canada is responsible for evaluating the safety of novel foods meant to be directly consumed by Canadians, such as fresh fruit or packaged foods. The relevant novelty of the trait in question is considered on a case-by-case basis. Novelty is not exclusive to genetically engineered plants, but all includes plants that undergo some form of modification that do not have a previous record of safe use in the Canadian environment.

If a plant is declared to be a PNT, it must be put through a rigorous set of risk-assessment and biosafety procedures before it is declared 'safe' for unconfined environmental release and commercialization (non-novel). If it is declared to be a PNT, it then moves through the regulatory system by meeting certain biosafety protocols (see Fig. 3.3). Biosafety testing is based on scientific information and appropriate data relative to the environmental risk of the PNT compared to its counterparts established in the Canadian environment. Testing must show that the PNT, once released, is not able to comingle with native species of a similar genotype (gene flow) or become a weed (weediness), must have its pest potential tested, must not be toxic to humans and/or animals, and must not pose a threat to biodiversity (CFIA, 2023, Directive 2009-09). Each stage of the regulatory system is dictated by regulations set out in Directives (DIR). The series of Directives dictate the requirements a pending PNT must meet to be approved for unconfined environmental release. The Directives are vital parts of Canada's product-based safety evaluation system for PNTs/novel foods (Clark & Phillips, 2013).

Figure 3.3 is a simplified decision-making tree for plants with novel traits/novel foods. As indicated in Fig. 3.3, Health Canada oversees the regulatory process for foods, with CFIA having responsibility for feeds or crops. In both cases, if the feed/crop/food is not considered novel no pre-market assessment is necessary, and the product proceeds to commercialization. For feed/crops considered novel, a series of steps include pre-submission meeting of relevant bodies, and pre-market assessments. For foods considered novel, Novel Food Guidelines apply, and pre-market submission assessments are required. If the plant with novel trait or the novel food does not satisfy risk and safety assessments, approval is not forthcoming. If risk and safety assessments are met, decisions are posted on the CFIA website (feed/crops), or Health Canada's website (food) and commercialization is allowed. Feed/crops undergo variety registration and are listed in the Canadian Variety Transparency Seeds Database prior to commercialization.

PNTs are subjected to a very complex and detailed risk assessment process. Health Canada also has rigorous thresholds for what is considered novel and how a novel food is treated in the regulatory system. Considering the rigor of the regulatory system in place to assess PNTs, it is significant that both Health Canada and CFIA declared that gene edited plants are not considered novel simply because they

---

[10] All plants that undergo some form of modification that have not previously been used in Canada could be considered novel, including gene edited agrifoods.

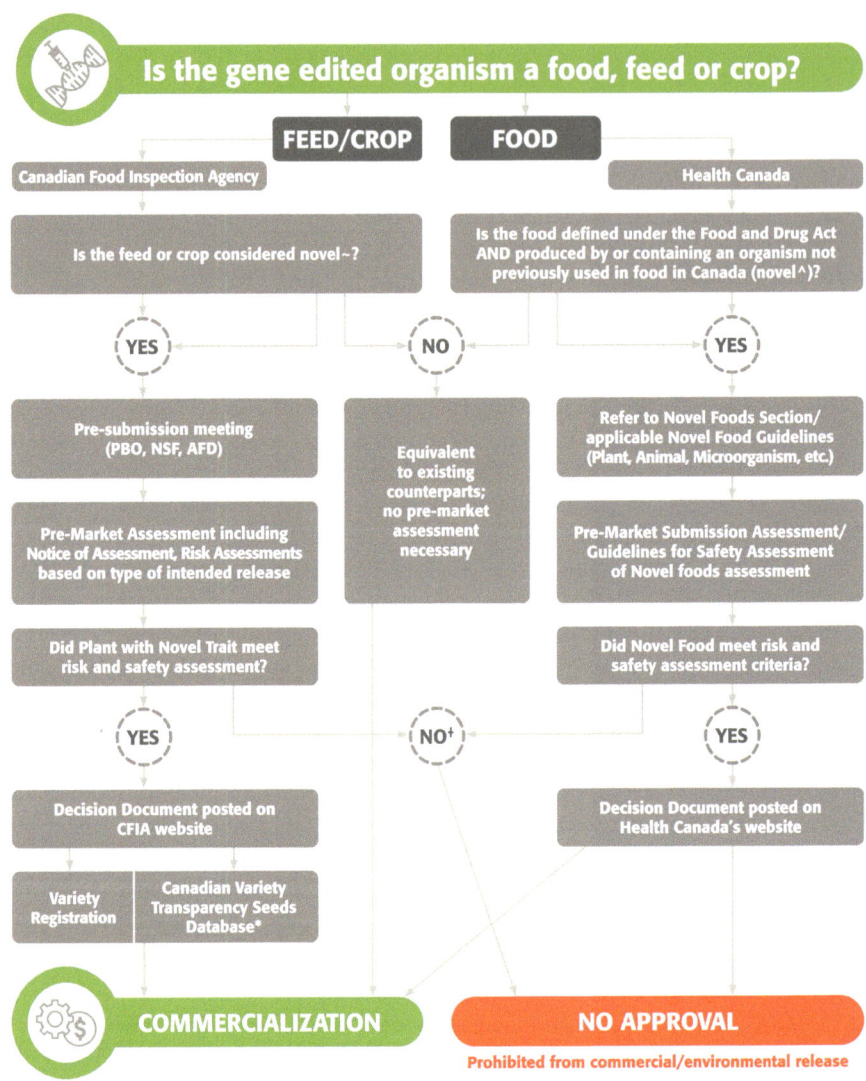

Fig. 3.3 Regulatory pathway for novel crops, foods and feeds in Canada

are derived from gene editing. All agrifoods, regardless of how they are modified, are subject to equal scrutiny in the Canadian regulatory system on a case-by-case basis based on scientific evidence.

Health Canada published a scientific opinion on the regulation of gene edited agrifood products. In 2022, Health Canada's Division 28 of the Food and Drug

Regulations (Novel Foods) was amended to include that novel food products that do not have a history of safe use are subject to evaluation regardless of technique. It stated that "novel food products from any breeding technique that might represent a food safety hazard would require a food safety assessment, to be done according to domestic guidance based on the Codex Guidance framework for safety assessments of foods derived from biotechnology" (FAO, 2022a: 32). This decision was preceded by Health Canada's scientific position in 2021 that determined a high amylopectin starch maize using an SDN1 technique was not considered a novel food product. It did not require pre-market safety assessment as a novel food. The rationale for this decision was that the product had the same phenotype as pre-existing commercial maize varieties with a similar spontaneous mutation and had a history of safe use as food and was therefore, not novel. The following year, a gene edited high oleic soybean was determined to be a novel food and was subject to a food safety assessment based on WHO/FAO expert consultations (FAO, 2022a: 32).

In May 2023, the CFIA announced its scientific position on gene edited agrifoods. In the updated Directive 2009-09 (Plants with novel traits registered under Part V of the Seeds Regulation: Guidelines for determining when to notify CFIA) Sect. 4.1 states:

> It is the scientific opinion of the CFIA that gene editing technologies do not present any unique or specifically identifiable environmental or human health safety concerns as compared to other technologies of plant development. For this reason, gene edited plants are regulated using a product based approach, like any other product of plant breeding. Namely, it is the traits that a plant exhibits and whether these traits would have a significant negative impact on environmental safety that are used to determine whether a plant would be subject to Part V of the *Seeds Regulations*. (www.inspection.gc.ca). (CFIA, 2023)

This decision was made based on government consultations with stakeholders throughout the agriculture sector and science-based assessment of the safety and efficacy of gene editing techniques. As Informant 15 (a regulatory representative) stated,

> …there's been a real need for larger conversations across the sector. Not necessarily to change how something's regulated, but to have an overall agricultural sector approach.

As of 2023, Canada is the only country to engage in multi-stakeholder consultations regarding the regulation and environmental release of gene edited agrifoods. In early 2023, the Government Technical Committee on Plant Breeding consulted with concerned stakeholders on potential issues that may arise with the introduction of gene edited seeds in the agrifood sector in Canada. The Committee focused on three primary areas of discussion: achieving transparency goals, establishing a governance structure to review effectiveness of transparency and government oversight, and discussing options for oversight of the Canadian Variety Transparency Database (AAFC, 2023).[11] When asked what issues arose during discussions among stakeholders, Informant 15 stated:

---

[11] This database is part of the broader seed industry led effort to provide varietal level transparency. As per the Heath Canada Transparency Initiative, where a variety has been developed using gene

the need for a transparent and predictable regulatory pathway for products and gene editing. How a particular policy links into the larger framework and context that surrounds it. Whether that be public perception around new technologies or tools required to enable coexistence. And by that, I mean the use of different technologies side by side.

Marketplace transparency was at the forefront of consultations, as noted in the *Chair's Report to the Minister on the Industry-Government Technical Committee on Plant Breeding Innovation Policy* (AAFC, 2023). The organic sector had concerns regarding how gene edited seeds will be identified in the agrifood system. The organic sector was concerned about how gene edited seeds would be segregated from conventional seeds (which the Canadian Organic Standards allows for use in organic agriculture if no organic varieties are available). Improved marketplace transparency in the sale of seed is paramount to ensure different types of agriculture can co-exist across Canada. Informant 8, a private sector representative who participated in these discussions surrounding the Canadian Variety Transparency Seeds Database, explained to us that:

> Between January and March [2023] there was one forum, and that resulted in a report from the AAFC[12]...And one of the outcomes of that was that report made a number of recommendations... It's a listing of registered crop variety in Canada, and it tells you whether they were developed using gene editing or not. One of their recommendations was that the database should remain free, and it should evolve over time and one of the recommendations was that AAFC should monitor and do some surveys of seed companies to make sure it's accurate, and up to date. And then another recommendation was that ... the ag sector and government continue to meet to monitor transparency as the first gene edited products are commercialized to make sure everybody's needs are getting met and that group has met twice now. So, this is not a popular opinion, but I think it's awesome. I never expected to be part of meetings where the organic sector sits down with tech developers, and they're having a conversation.

Stakeholders such as CropLife Canada and Seeds Canada[13] agreed to participate in the building and continued updating of the Canadian Variety Transparency Database (the 'Seeds Database') and Health Canada's Transparency Initiative. The Seeds Database will be free, publicly accessible, regularly updated, reliable and user-friendly. The Technical Committee proposed that AAFC would be responsible for regular audits and conduct surveys with seed developers to confirm they are using

---

editing technology and does not meet criteria of 'novel food', it will be required to appear under "Health Canada Notification" column of the database. More information on these varieties can be found on the 'non-novel' list'. Varietal data is collected from CFIA records, Seeds Canada members and partners. For more information see Seeds Canada (2023): https://seeds-canada.ca/en/seed-resources/transparency-database/

[12] Agriculture and Agri-Food Canada (the Canadian federal government ministry of agriculture).

[13] CropLife Canada is an organization that represents Canadian manufacturers, developers and distributors of pest control and modern plant breeding products (https://croplife.ca/about-us/). Seeds Canada is the amalgamation of four organizations: Canada Plant Technology Agency, the Commercial Seed Analysis Association of Canada, the Canadian Seed Institute and Canadian Seed Trade Association (https://seeds-canada.ca/en/about).

the Seeds Database.[14] Though this Database is not part of the regulatory scheme, stakeholders have committed to its creation and maintenance as a voluntary condition that needs to be met for co-existence to be possible in the Canadian agrifood sector and for trust in the regulatory system to be maintained. Seeds Canada is responsible for the Database's operation and implementing improvements. Informant 15 said that the Canadian grain sector has much to gain from the Seeds Database as well as the organics sector:

> I am not aware of any other country from which the seeds sector has put forward a voluntary transparency database. I think it's unique to Canada…the need has come up through the organic sector looking to make sure that the seeds they buy are eligible for organic certification. But really, it's also been a need identified by the grain sector as well.

Unlike in other jurisdictions, where the politics surrounding agricultural biotechnology has stalled or in some cases prevented the conversations from ever happening, transparency and deliberative discussions are the cornerstone of the Canadian model to increase stakeholder access to all the modern benefits of agricultural science, while avoiding imposing unnecessary costs on non-users. This approach is not without its challenges but has prioritized the economic health of the agricultural value chain in general above whether one approach to agriculture is considered superior to the other. All approaches to agrifood production have their benefits and drawbacks.

This is not to say that all stakeholders agree with how Health Canada, CFIA and AAFC have proceeded with regulating gene edited agrifoods. There is plenty of dissent. For example, representatives of the organic sector stated that transparency measures, in terms of gene edited seeds and the Seeds Database, do not go far enough to protect organic certification. The active participation of the organic sector in the proposed governance structures and oversight activities and ongoing consultations with plant breeders will hopefully live up to the expectations of all stakeholders in the agriculture sector (AAFC, 2023). But from a policy perspective, the path that the regulatory agencies have taken is one based on transparency, responsiveness, the principles of governance and above all, science-based risk assessments.

The result of the stakeholder consultations has helped inform the regulatory pathway for gene edited agrifoods in Canada. Some products developed using gene editing techniques may not meet the regulatory definition of 'novel'. If not, a product is considered equivalent to its existing counterparts, and no pre-market assessment is required. Health Canada and CFIA have declared that gene edited agrifoods are not considered novel, and therefore not subject to the regulatory process for PNTs. This is the first step towards freeing gene edited foods from unnecessary regulatory burden, allowing for useful advancements in plant breeding and nutritional profiles of foods to be utilized, and getting us closer to solving some of the current supply chain problems plaguing the global agrifood system.

---

[14] It was also suggested that if it is determined that seed developers are not disclosing information about gene edited varieties in the Database, additional measures may need to be considered.

3.5   Conclusion

As new breeding techniques emerge, some are asking if new breeding techniques that do not include foreign DNA (SDN1) should be treated differently than GMOs or considered equivalent (in terms of risk assessment) as agrifoods from traditional breeding techniques. There are governance issues and questions regarding current and future novel breeding techniques and what types of consultation should be used. Canada has embraced the deliberative governance approach to gene edited agricultural plants. It has considered numerous stakeholders and their preferences, while trying to create a transparent regulatory framework. Risk and innovation need to be balanced to provide opportunity and benefits to all stakeholders—businesses and citizens. But risks to human, plant, animal, or environmental health must be given priority. How the level of risk is determined is what is often at odds in different countries. Moving forward, what appears to be central to effective regulation of commercialized gene edited agrifoods is working on communication exchanges among all stakeholders regarding policy objectives that balance precaution with facilitating innovative gene editing agrifoods.

One of the challenges of rapid advancement of techniques and gene edited organisms is that regulatory systems may struggle to keep up with the demand for regulatory approval. This may overwhelm understaffed and under resourced regulatory systems, compromising efforts to be transparent and inclusive in decision making and risk analysis. "Science cannot settle normative questions or determine policy judgements and decisions about regulating genetic engineering merely on its own assumptions, as both values and interests jointly contribute to framing social choices about the data to be acquired, analyzed and interpreted" (Leone, 2019).

Informant 5 (a research scientist) shares a similar perspective. The knowledge deficit model approach to deliberative governance and innovative technologies (manufactured risks) cannot make definitive conclusions on normative values applied to science. Informant 5's position is that:

> I'm not one of those people [who] believes that you have to convince people…you have to get them to accept the science, to accept the judgment. What they need to do is have some faith and trust and understanding that that the things that matter are dealt with…Nobody has to agree on the science. They just have to say, this is a legitimate and appropriate policy. Science never tells you what to do. Science just gives you the choices and the costs and benefits, and the trade offs.

This not only applies to gene edited agrifoods, but several innovative technologies such as vaccines that garnered vigorous public debate about the legitimacy of the science behind them.

References

Agriculture and Agrifood Canada (AAFC). (2023). *Chair's report to the minister on the industry-government technical committee on plant breeding innovation transparency.* Government of Canada. Retrieved January 14, 2024, from https://agriculture.canada.ca/en/department/transparency/chairs-report-minister-industry-government-technical-committee-plant-breeding-innovation

Ahmad, A., Munawar, N., Khan, Z., Qusmani, A. T., Khan, S. H., Jamil, A., Ashraf, S., Ghouri, M. Z., Aslam, S., Mubarik, M. S., et al. (2021). An outlook on global regulatory landscape for genome-edited crops. *International Journal of Molecular Sciences, 22*(21), 11753. https://doi.org/10.3390/ijms222111753

Asquer, A., & Krachkovskaya, A. (2021). Uncertainty, institutions and regulatory responses to emerging technologies: CRISPR gene editing in the US and the EU (2012–2019). *Regulation and Governance, 15*, 1111–1127.

Canadian Food Inspection Agency (CFIA). (2023). *Directive 2009-09: Plants with novel traits regulated under Part V of the seeds regulations: Guidelines for determining when to notify the CFIA.* Retrieved October 14, 2023 from https://inspection.canada.ca/plant-varieties/plants-with-novel-traits/applicants/directive-2009-09/eng/1304466419931/1304466812439#a4_1

Chou, F. L. (2023, November 7). Innovation in gene editing and plant breeding, American Seed Trade Association. Farm Foundation Forum. In *Innovation in gene editing and plant breeding: A look at scientific advancement and consumer perspectives in food and agriculture.* Retrieved December 2, 2023, from https://www.youtube.com/watch?v=5XB02AOzB7I&list=PLqEx8zAfSCyLkRKlG1g6aq3CU4Uvm23mY

Clark, L. F., & Phillips, P. W. B. (2013). Bioproduct approval regulation: An analysis of front-line governance complexity. *AgBioForum, 16*(2), 112–125.

Convention on Biological Diversity. (2023). *Frequently asked questions on the Cartagena protocol.* Retrieved December 15, 2023, from https://bch.cbd.int/protocol/cpb_faq.shtml#faq3

Convention on Biological Diversity (CBD). (2018). *Parties to the Cartagena protocol and its supplementary protocol on liability and redress.* Retrieved October 12, 2023, from https://bch.cbd.int/protocol/parties/

Dobrovidova, O. (2019). Russia joins in global gene-editing bonanza. *Nature, 569*(7756), 319–320. https://doi.org/10.1038/d41586-019-01519-6

Entine, J., Felipe, M. S. S., Groenewald, J. H., et al. (2021). Regulatory approaches for genome edited agricultural plants in select countries and jurisdictions around the world. *Transgenic Research, 30*, 551–584. https://doi.org/10.1007/s11248-021-00257-8

Eriksson, D., Kershen, D., Nepomuceno, A., Pogson, B., Prieto, H., Purnhagen, K., Smyth, S., Wesseler, J., & Whelan, A. (2019). A comparison of the EU regulatory approach to directed mutagenesis with that of other jurisdictions, consequences for international trade and potential steps forward. *New Phytologist, 222*(4), 1673–1684.

FAO (Food and Agriculture Organisation). (2022a). *Gene editing and agrifood systems.* FAO (Food and Agriculture Organisation). Retrieved January 3, 2024, from https://doi.org/10.4060/cc3579en

FAO (Food and Agriculture Organisation). (2022b). *FAO GM Foods Platform. Browse information by country.* Retrieved October 12, 2023, from https://www.fao.org/food/food-safety-quality/gm-foods-platform/browse-information-by-country/en/

Friedrichs, S., Takasu, Y., Kearns, P., Dagallier, B., Oshima, R., Schofield, J., & Moreddu, C. (2019). An overview of regulatory approaches to genome editing in agriculture (Conference report). *Biotechnology Research and Innovation, 3*, 208–220.

Gatica-Arias, A. (2020). The regulatory current status of plant breeding technologies in some Latin American and Caribbean countries. *Plant, Cell, Tissue and Organ Culture, 141*, 229–242.

Genetic Literacy Project. (2023). *Gene editing regulation tracker.* Retrieved December 13, 2023, from https://crispr-gene-editing-regs-tracker.geneticliteracyproject.org/

Genetic Literacy Project. (2024a). *Bangladesh Greenlights gene editing to 'meet the needs of farmers and consumers'*. Retrieved March 12, 2024, from https://geneticliteracyproject.org/2024/03/12/bangladesh-greenlights-gene-editing-to-meet-the-needs-of-farmers-and-consumers/

Genetic Literacy Project. (2024b). *Norway: Crops and food*. Retrieved March 12, 2024, from https://crispr-gene-editing-regs-tracker.geneticliteracyproject.org/norway-crops-food/

Gould, F., Amasino, R. M., Brossard, D., Buell, C. R., Dixon, R. A., Falck-Zepeda, J. B., Gallo, M. A., et al. (2022). Toward product-based regulation of crops. *Science, 377*(6610), 1051–1053. https://doi.org/10.1126/science.abo3034

Jones, M. G. K., et al. (2022). Enabling Trade in Gene-Edited Produce in Asia and Australasia: The developing regulatory landscape and future perspectives. *Plants, 11*(2538), 1–36.

Laaninen, T. (2020). *New plant-breeding techniques: Applicability of EU GMO rules*. European Parliamentary Research Service. Retrieved October 13, 2023, from https://www.europarl.europa.eu/RegData/etudes/BRIE/2020/659343/EPRS_BRI(2020)659343_EN.pdf

Lassoued, R., Phillips, P. W. B., Macall, D. M., Hesseln, H., & Smyth, S. J. (2021). Expert opinions on the regulation of plant genome editing. *Plant Biotechnology Journal, 14*(4), 321–337. https://doi.org/10.1111/pbi.13597

Lawrence, R. (2023, November 7). *Gene editing as a tool to create targeted variation in plants*. Farm Foundation Forum, Innovation in Gene Editing and Plant Breeding: A look at scientific advancement and consumer perspectives in food and agriculture. Retrieved December 2, 2023, from https://www.youtube.com/watch?v=5XB02AOzB7I&list=PLqEx8zAfSCyLkRKlG1g6aq3CU4Uvm23mY

Leone, L. (2019). Gene editing for the EU agrifood: Risks and promises in science regulation. *European Journal of Risk Regulation, 10*, 766–780. https://doi.org/10.1017/err.2019.55

Menz, J., Modrzejewski, D., Hartung, F., Wilhelm, R., & Sprink, T. (2020). Genome edited crops touch the market: A view on the global development and regulatory environment. *Frontiers in Plant Science, 11*, 586027. https://doi.org/10.3389/fpls.2020.586027

Modrzejewski, D., Hartung, F., Sprink, T., Krause, D., Kohl, C., & Wilhelm, R. (2019). What is the available evidence for the range of applications of genome-editing as a new tool for plant trait modification and the potential occurrence of associated off-target effects: A systematic map. *Environmental Evidence, 8*, 27. https://doi.org/10.1186/s13750-019-0171-5

NASEM (National Academies of Science Engineering and Medicine). (2017). *Preparing for future products of biotechnology*. NASEM (National Academies of Science Engineering and Medicine). https://doi.org/10.17226/24605, https://www.ncbi.nlm.nih.gov/books/NBK442207/

Organisation for Economic Co-operation and Development (OECD). (2018). *Conference on genome editing: Applications in agriculture*. Retrieved on December 2, 2023, from: https://www.oecd.org/environment/genome-editing-agriculture/

Prakash, C. S. et al. (2014). *Scientists in support of agricultural biotechnology. 2000–2014*. Agbioworld. Retrieved on October 17, 2023, from: www.agbioworld.org/declaration/index.html

Qaim, M. (2009). The economics of genetically modified crops. *Annual Review of Resource Economics, 1*(1), 665–694. https://www.annualreviews.org/doi/10.1146/annurev.resource.050708.144203

Seeds Canada. (2023). *Canadian variety transparency database*. Retrieved on December 21, 2023, from: https://seeds-canada.ca/en/seed-resources/transparency-database/

Shukla-Jones, A., Friedrichs, S., & Winickoff, D. (2018). *Gene editing in an international context: Scientific, economic and social issues across sectors* (OECD Science, Technology and Industry Working Papers, No. 2018/04). OECD Publishing. Retrieved on April 13, 2023, from: https://doi.org/10.1787/38a54acb-en

South Asia Biosafety Program. (2021). *Genetically engineered plants and biosafety*. Retrieved June 10, 2023, from: https://bangladeshbiosafety.org/wp-content/uploads/2021/03/02_FAQ_GE_Plants_Biosafety_web.pdf

Sprink, T., Eriksson, D., Schiemann, J., & Hartung, F. (2016). Regulatory hurdles for genome editing: Process- vs. product-based approaches in different regulatory contexts. *Plant Cell Reports, 35*(7), 1493–1506. https://doi.org/10.1007/S00299-016-1990-2

Tagliabue, G. (2017). Product, not process! Explaining a basic concept in agricultural biotechnologies and food safety. *Life Sciences, Society and Policy, 13*(1), 3. https://doi.org/10.1186/s40504-017-0048-8

Tiwari, M., Trivedi, P. K., & Pandey, A. (2021). Emerging tools and paradigm shift of gene editing in cereals, fruits and horticultural crops for enhancing nutritional value and food security. *Food and Energy Security, 10*(1), e258. https://doi.org/10.1002/fes3.258

United States Department of Agriculture (USDA). (2019, June 5). *GAIN report: Soybeans Bolivia adopts biotechnology*. Retrieved on June 13, 2023, from: https://apps.fas.usda.gov/newgain-api/api/report/downloadreportbyfilename?filename=Soybeans%20Bolivia%20Adopts%20Biotechnology_Lima_Bolivia_6-5-2019.pdf

United States Department of Agriculture (USDA). (2024, February 27). *GAIN report: Costa Rica opens door to innovative biotechnologies*. Retrieved on March 17, 2024, from: https://fas.usda.gov/data/costa-rica-costa-rica-opens-door-innovative-biotechnologies

Vora, Z., Pandya, J., Sangh, C., & Vaikuntapu, P. R. (2023). The evolving landscape of global regulations on genome-edited crops. *Journal of Plant Biochemistry and Biotechnology, 32*(4), 831–845.

Whelan, A. I., & Lema, M. A. (2019). Regulation of genome editing in plant biotechnology: Argentina. In H. G. Dederer & D. Hamburger (Eds.), *Regulation of genome editing in plant biotechnology* (pp. 19–62). Springer. https://doi.org/10.1007/978-3-030-17119-3_2

World Trade Organisation (WTO). (1994, April 15). *TRIPS: Agreement on trade-related aspects of intellectual property rights*. Marrakesh Agreement Establishing the World Trade Organization, Annex 1C, 1869 U.N.T.S. 299, 33 I.L.M. 1197.

Part II
Emerging Opportunities for Regulatory Enhancement

Chapter 4
Novel Genomic Techniques and Applications on the Horizon

4.1 Introduction

The European Union in July 2023 released a proposal entitled, 'Proposal for a REGULATION OF THE EUROPEAN PARLIAMENT AND OF THE COUNCIL on plants obtained by certain new genomic techniques and their food and feed, and amending Regulation (EU) 2017/625' (EFSA, 2023). The proposal states,

> The European Food Safety Authority (EFSA) concluded that, as regards risks for human and animal health and the environment, there are no specific hazards linked to targeted mutagenesis or cisgenesis…Climate change and biodiversity loss have put the focus on long-term resilience of the food chain and the need to transition to more sustainable agriculture and food systems. The European Green Deal's Farm to Fork Strategy specifically identifies new techniques, including biotechnology, that are safe for consumers and the environment and bring benefits to society as a whole, as a possible tool to increase sustainability of agri-food systems and contribute to guaranteeing food security (EFSA, 2023: 2).

This is a significant announcement, as it signals a major shift from the precautionary approach embedded in how the EU regulates products of biotechnology in agriculture. The catch-all terminology of 'new breeding techniques' used in the proposal covers technologies in the pipeline and future technologies that can be applied to agrifood. The motivation behind this proposal is to work towards a sustainable food system within the European Union and to meet current and future food security goals in the face of climate change and economic uncertainties. This chapter examines a sampling of new breeding techniques in the pipeline and how they are enabling other technologies and gene editing platforms. Though our primary focus is agrifood plants, we draw attention to gene editing in related research areas, such as microbes, which are components of the broader ecosystem and fundamental to sustainable agrifood systems. This chapter focuses on the category of 'open release products' (organisms in the form of seeds, and plants for human/animal consumption released into the environment), as they are most relevant to agrifood

L. F. Clark, J. E. Hobbs, *International Regulation of Gene Editing Technologies in Crops*, SpringerBriefs in Environmental Science, https://doi.org/10.1007/978-3-031-63917-3_4

production. We take a deeper look at current research being conducted on orphan crops,[1] and re-wilding or 'wide crosses' of crops that are important agronomically and for food security.

According to the 2017 report by National Academies of Science, Engineering and Medicine (NASEM), three classes of future gene edited products are under development: platforms, contained products, and open release products (NASEM, 2017). Platforms are tools used to create other biotech products. Contained products encompass organisms in contained environments like labs. Open release products include plants, animals, microbes, and synthetic organisms that are designed to be released into the environment. We discuss each of these classes of future gene edited products.

4.2 Future Gene Editing Technologies: Platforms

Platforms for biotechnology—tools used in the creation of other biotech products—include 'wet lab' products such as enzymes, vectors, cells, sequencing prep kits and 'dry lab' products like computer software (e.g., Genome Analysis Toolkit). In this vein, precision breeding or 'smart' breeding draws from engineering, molecular biology, agricultural science, and computer science (bioinformatics) and relies heavily on dry lab products. Generally, it uses Artificial Intelligence (AI) and automation & sequencing platforms to guide genetic changes so that scientists can quickly analyze changes to assess valuable traits and possibly remove negative plant traits using data science tools. Precision breeding can tailor traits to producers' needs, which may differ across geographic regions, climates, or soil compositions.

Having widely accessible databases of genome sequence data can help develop AI models that can efficiently and accurately predict how plants edited for agronomic traits will fare under particular biotic and abiotic stresses. For example, Newman and Furbank (2021) used the Australian National Variety Trial (ANVT) database to develop integrated machine learning to accurately predict yields and agronomic traits. The ANVT is the largest independent coordinated national field trial network in the world. There are more than 650 trials sown across 300 locations across 10 species types including barley, canola, lentil, oats, and wheat (Grain Research Development Corporation, 2023). Machine learning allows for compressive analysis of large amounts of data, such as that in the ANVT to retrieve performance-based metrics quicker than by using previous techniques (Zhang et al., 2023).

Organisms in contained environments like labs, include microbes or are synthetic based, rather than an animal or plant host. Examples include gene edited microbes in fermenters to produce chemicals, fuels, polymers, or food additives.

---

[1] Orphan crops are staple crops namely grown in Africa, Asia and South America for domestic consumption. They are not typically part of global agrifood trade.

Using gene editing on microbes is a strong area of interest (NASEM, 2017). Research is underway to discovery how gene editing can be used to develop micro-organisms to control pests, weeds, improve soil quality and aid in food processing (Wesseler et al., 2022). Scientists are looking at microbes such as yeast, algae, and bacteria to produce useful gene edited chemical biofuels. For example, the International Service for the Acquisition of Agri-biotech Applications (ISAAA) reports that the *Nannochloropsis* species of algae accumulates large amounts of lipids through photosynthesis and these lipids have the potential to be used as feed-stocks or biodiesels. Currently, the technology needs to evolve to make biodiesel from algae cost-effective on a global scale.

Examples of the application of organisms in contained environments include developing resistance in microorganisms to specific plant pathogens, microbes that improve nitrogen fixation in the soil, and alternatives to pesticides, herbicides and fungicides in agricultural production. Novel bacteria and enzymes can be used in food processing in the fermentation process, to develop plant-based proteins, and provide alternatives to fossil fuels such as biochemicals (FAO, 2022: 12). These could be important contributions to the Farm to Fork strategy in the EU and help improve soil quality in areas with depleted soil and dependency on marginal lands for subsistence. One future expectation is using gene editing to insert synthetic DNA into genomic sequences of microbes. For example, these may help improve nitrogen fixation in the soil and assist with bioremediation in contaminated soil sites (NASEM, 2017: 23).

Several of the research scientists we interviewed expressed excitement about gene editing applications on the horizon, especially gene editing's applications to microbes. As more knowledge is accumulated about a specific genome, the abilities for gene editing to offer solutions to some of today's problems continues to grow. As Informant 4 (an academic researcher) stated,

> …[gene editing technology] is one of the many tools that researchers in the scientific community have access to and it's really going to change and it's already changing the way we conduct science, because it's combined with all the genomic information we have about the plants of interest…but we are getting more genomic information about related crop species. So not just those that have been domesticated. It is going to have a significant impact on how we are going to develop the next crops, the next food, and some scientists are thinking of using genome editing also on microbials to moderate the composition of the microbiome. These tools are opening up new ways of conducting science, new types of products that could be developed because obviously, you can expand the genetic diversity to what is currently available.

Another informant, who is a research scientist working for an international organization, is excited about the applications of gene editing to microbes and how it can help reduce agrochemical use in food production. Informant 18 tells us,

> [With] microbes, there's a common phenomenon which is known as 'quorum sensing'… Which is when one microbe communicates with another through sensing, to do the similar thing. It's like a telephone or mobile phone or Twitter if you may…There is a process where nematodes are told through quorum sensing that there is no plant to infect. Can you beat that? So there's a plant, there are nematodes, but they don't go and infect this host plant

because to them, there is no plant. So that means less or no pesticide spray. And nematodes are happy. The plant is happy. It's feeding on something else, but not on your host plant.

There are remarkable innovations in the pipeline with gene editing that could revolutionize the agrifood system and make positive contributions to global food security. CRISPR-Cas9 is widely used to make these discoveries and make applications precise. However, there are other gene editing techniques on the horizon that use proteins other than Cas9 to edit the genome.

Open release products include plants, animals, microbes, and synthetic organisms that are designed to be released into the environment. The ability for gene edited organisms to exist without human intervention is cited as the major distinction between gene edited products in the past, and potentially, those in the future. The types of environments where these organisms will be released also varies. For example, some gene edited plants may be designed to exist in forests, pastures, or cityscapes, while microbes may exist underground in mines, waterways, in the soil, and in animal digestive tracts. Experts at the NASEM committee meeting thought that most of the products under development would be directed toward agrifood production, but also believed that they would be used for soil decontamination, and 'lab' meat derived from animal cells (NASEM, 2017).

Several novel open release applications of genomic techniques emerging from labs may create future regulatory challenges. Gene drive systems, often based on CRISPR-Cas9, allow an edited gene on a chromosome to copy itself onto its partner chromosome during cell division. Because the edited gene is copied on the partnered chromosomes, inheritance is 100% rather than 50%. 'Cargo' genes are required for a gene drive to work (Coffey, 2020). Cargo transgenes can be designed to confer chosen traits. Using CRISPR-Cas9 to introduce 'cargo genes' confers any trait that can be genetically linked to an engineered drive system. The genetic trait engineered via the drive can spread through a population. Gene drives alter an organism's genes in a way that ensures that all offspring take on the edited traits, enabling a rapid spread of engineered genes through normal reproductive channels.

Scientists are experimenting with gene drives to control populations of invasive species or disease-carrying insects, such as the malaria-carrying mosquito of the *Anopheles* genus. Gene drives may also be gamechangers in terms of crop breeding (pest management, invasive species management). Research is under way to investigate potential applications to control agricultural weeds using these techniques (Neve, 2018). Over time, this could eliminate pest populations or decrease unwanted species in wild ecosystems (FAO, 2022). Researchers are working on developing gene drives that are limited to specific geographical regions that would help eradicate agricultural pests but are also adding immunization genes that could protect valuable and vulnerable populations of organisms (Bier, 2022).

Despite the promise of gene drives, there are concerns about unintentional consequences of using gene drives to genetically edit species for unconfined environmental release (Webber et al., 2015). Hayes et al. (2018) discusses potential hazards of open release gene drives into the environment. Risks include the introduction of gene drive organisms into non-target, related species in ecosystems. Hybridization or horizontal gene transfer could also have negative impacts on ecosystems. Nevertheless, to date there is little evidence of successful gene drive cases in plants.

Informant 12 (an academic researcher) spoke at-length about gene drives and why they are currently not a large part of the conversation on biotechnology in the agrifood system. They said,

> there are a couple of different concerns. One is unrestricted spread. Usually, we want to control invasive pests and not native pests. And so native versus non-native is also remarkably important to ecologists. If you release a gene drive, that is a suppression drive that is really pushing down that pressure on replacement. That inoculates the insect, for example, from being able to host a certain pathogen in it. It could spread throughout native areas as well as the invasive areas. It's very hard to control this kind of stuff. That's how the invasive species got there in the first place. So technologically, are we there to control that spread? ...there's a trend towards more restricted and controlled use of gene drives.

While there remain questions about the ability to control gene drives once they have been released and their potential applications for benefits to agrifood production, there are other gene edited technologies in the pipeline that carry less uncertainty regarding unintended consequences.

4.3 Climate Change and Biodiversity

Gene editing offers potential solutions to help address agricultural challenges that have accompanied climate change. Plant domestication and genetic improvement have been important contributors to current levels of agricultural productivity and the global food supply. Improved crop yields ushered in by the first Green Revolution were achieved by creating high-yielding, lodging-resistant, fertilizer-responsive varieties (Vikram et al., 2015). The traits preferred during the domestication process for cereals for example, favoured a reduction in seed shattering, and absence of secondary dormancy,[2] while vegetable crops were bred for specific fruit size and shape, pigmentation, ease of planting/harvesting and transportation (Atwell et al., 2014; Fernie & Yan, 2019). However, many of the inherited traits involved in biotic and abiotic stress resistance may have been weakened or lost during this process of domestication for selected agronomic traits. While these domesticated crops perform well under ideal (temperature and soil moisture) conditions, these varieties are not necessarily suited to perform well under extreme climate conditions or on marginal soils. It has imposed limitations on the environments in which these crops can be efficiently cultivated. As climate change advances at a rapid pace, farmers are increasingly less able to rely on the 'stable' environmental conditions that these staple crops were bred to grow in.

---

[2] An evolutionary adaptation that prevents seeds from germinating in unfavourable ecological conditions, e.g., droughts, floods after ripening. Secondary dormancy may contribute to weediness.

Agriculture is extremely vulnerable to climate change. Many crops the world depends on were bred for temperature and precipitation ranges that are far less predictable today than previously, though farmers have always had to deal with extreme weather events such as droughts and floods, as well as pests. As Razzaq et al. (2021: 6124) state, climate change brings more frequent extreme weather events, which may lower long-term yields by damaging crops at various stages of development (Moriondo et al., 2010; Porter et al., 2014). As a result, the timing of field applications of fertilizers or pesticides is more difficult to predict, which can also contribute to decreasing yields (Antle et al., 2004; Tubiello et al., 2007). Changes in temperature, reductions in rainfall in certain areas and an over-abundance of precipitation elsewhere, influences the lifecycles of pests and negatively impacts soil composition. The quality of products will also be impaired, as elevated CO_2 contributes to a reduction in protein content in cereal grains (Sinclair et al., 2000; Gornall et al., 2010). All these factors have put continued stress on the entire global agrifood supply chain, while contributing to growing global food insecurity.

The current progress in agronomy and crop breeding is not sufficient to keep up with the required increase in food production, prompting a need for a major shift in the breeding paradigm to create stress-resilient crops. The increasing efficiency and affordability of genome sequencing, and the development of novel genome editing tools opens new and exciting prospects for harnessing the potential of climate-resilient crop varieties, and investigating the genes present in wild relatives lost during the domestication process (Razzaq et al., 2021: 6133).

As such, researchers around the world are investigating how gene editing can help improve the resilience of 'orphan crops'[3] that have long been neglected. So-called orphan crops are those which are not a major focus of crop breeding or international trade in agricultural commodities but may have a role to play in specific regions. There is also research underway regarding the potential of 're-wilding' domesticated staple crops, like wheat and rice with traits from wild relatives that have been lost through domestication. But as climate change predictions point to warmer temperatures on the horizon, changes to rainfall patterns, and increased frequency and severity of extreme weather, the urgency to find ways of mitigating the effects of climate change on the world's food supply have become of the utmost importance (Wheeler & von Braun, 2013; Zaid et al., 2020).

---

[3] "Orphan (or minor) crops are those crops which are typically not traded internationally but which can play an important role in regional food security. For various reasons, many of these crops have received little attention from crop breeders or other research institutions wishing to improve their productivity." https://fse.fsi.stanford.edu/research/orphan_crops#:~:text=Orphan%20(or%20 minor)%20crops%20are,wishing%20to%20improve%20their%20productivity; also, Naylor et al. (2004).

4.3.1 Orphan Crops

Orphan crops go by different names in the literature, including 'underutilized', 'minor', 'neglected', 'promising', 'niche' and/or 'traditional' (Yaqoob et al., 2023: 1). The category of orphan crops includes plants such as buckwheat, quinoa, cassava, banana, pigeon pea, millets, and many others. These crops are grown around the world by smallholder farmers in Latin America, Africa, and Asia. These crops are often resilient, can grow on marginal land, in lower quality soil and in some cases are more able to endure biotic and abiotic stresses than staple crops that have had some traits (beneficial to changing temperature and precipitation conditions) bred out of them to maximize yields in intensive monocropping. Some of the traits of orphaned crops make them less economical to farm on an intensive scale, such as smaller fruit size, low yield and less than ideal plant structure (Lyzenga et al., 2021: 142). We were curious to get an insight into how crops are chosen for genomic research, and eventually commercialization. In response to the question, 'how are crops chosen for research purposes?' Informant 17 (a research scientist) replied,

> it's acreage rate. And I'm not sure if 'popularity' is the word, but…There's not a lot of external funding for [small acreage crops].

Because of the lack of their commercial importance, orphan crops have historically received less research funding and genomic resources than their staple counterparts.

Even though orphan crops have not been priority crops for research and investment, they hold a lot of potential to meet UN Sustainable Development Goals (e.g., zero hunger) in lower-income countries across Africa, Asia, and Latin America. Gene editing can be used to improve nutritional profiles of orphan crops and improve agrobiodiversity (FAO, 2022: 14). The FAO (2022) cites a number of organizations that are prioritizing genetic research into orphan crops for the public good, including CGIAR (formerly the Consultative Group on International Agricultural Research), as well as AIRCA centres (Association of International Research and Development Centers for Agriculture) such as ICBA (International Center for Biosaline Agriculture) and ICIPE (International Centre for Insect Physiology and Ecology), and NARS (National Agricultural Research Systems). The Gates Foundation is also conducting research into the potential of gene editing to enhance orphan crops like cassava (FAO, 2022: 14).

In response to the question, 'How can gene editing be used to protect biodiversity and fight climate change? Are you a part of/are you aware of current research on gene-editing orphan crops?', Informant 17 (a research scientist) responded:

> …about the funding that rice, wheat or soybean would get, you would not expect it to be the same sustained effort that you would see for these millets or…legumes which are nutritious, more sustainable because I think they're not put across that way or the focus is not there. But if you're looking for something that is good for people, sustainable in the long run, we will have to find crops which fit that…and look for diverse crops, not just these three, four crops.

Wild relatives of orphaned crops may hold useful genetic traits that could improve agronomic outputs and nutritional profiles not found in cultivated crops. Informant 18, who works for an international development NGO, said that their organization's 'mandate crops' are "chickpea, pigeon pea, ground nuts, sorghum and finger millet." Some NGOs are researching how to use crops like pigeon pea to increase the productivity of marginal land. For example, Informant 18 is conducting research on what crops can be useful to intercropping strategies in India. As they explain,

> we are also looking for plant architecture …for increasing yields in the case of pigeon pea especially. It should fit intercropping. What happens in many parts of India after they harvest rice, there is a very small window in which the field remains fallow. If that can be brought into cultivation with short duration pulses, you are taking a crop which would otherwise not be taken and providing diverse food on the plate as well.

Many other orphan crops offer potential. Table 4.1 below, adapted from Yaqoob et al. (2023), shows the type of crop and the trait(s) that may be useful as climate pressures challenge the outputs of higher profile crops like maize, wheat, rice, and

Table 4.1 Major orphan crops and potential agronomic/food security characteristics

Orphan crop	Characteristic
African rice	Stress tolerance
Amarath	Nutrition
Bambara groundnut	Nutrition/drought tolerance
Barnyard millet	Abiotic stress tolerance
Buckwheat	Nutrition
Cassava	Drought tolerance
Chickpea	Nutrition
Cowpea	Nutrition/drought tolerance
Enset	Drought tolerance
Foxtail millet	Abiotic stress tolerance
Grass pea	Nutrition/extreme drought tolerance
Horsegram	Nutrition
Kodo millet	Abiotic stress tolerance
Lentil	Nutrition
Linseed	Nutrition
Little millet	Abiotic stress tolerance
Okra	Nutrition/Biotic stress tolerance
Pearl millet	Abiotic stress tolerance
Pigeon pea	Nutrition
Proso-millet	Abiotic stress tolerance
Quinoa	Nutrition
Sweet potato	Nutrition
Tef	Gluten-free/Abiotic stress tolerance
Yam	Drought tolerance

Adapted from Yaqoob et al., 2023: 3, licensed under CC-BY 4.0

soybean. Genomic research is underway to determine how to improve the agronomic profiles of these types of crops that have promise for global food insecurity.

Research is also underway to identify beneficial traits in wild relatives of orphan crops to improve agronomic profiles. *De novo* domestication has emerged as a potential way to harness the beneficial traits from wild species through gene editing and molecular breeding to create new, more climate resilient agrifoods. As Lyzenga (2021: 142) explains, many traditionally breeding techniques are ideal candidates for CRISPR-Cas gene editing platforms. However, challenges remain. There is a relatively high cost attached to phenotyping. The quality and control of seed systems and other crop inputs are other considerations in orphan crop research (FAO, 2022: 22). Without reliable seed distribution systems that farmers can depend on, the development of orphan crops, especially for smallholders in the developing world is much less impactful. The quality and control of seed systems and other inputs for crop production will play an important role in genomic research into orphan crops, as well as the pressing need to diversify agricultural production under the constantly changing environmental conditions of climate change.

4.3.2 Re-wilding

Re-wilding or 'wide crossing' involves taking traits from wild cousins of domesticated crops and using gene editing to improve genetic variation and biodiversity. Its goal is to reintroduce mutations into the cultivated crop gene pool that are still available in their wild relatives. The process of natural selection of wild relatives of some domesticated crops has resulted in an accumulation of genes that, for example, provide tolerance against pests, diseases, extreme temperatures, flooding, drought, and salinity (Montenegro et al., 2017). According to Cardi et al. (2023: 16), re-wilding can be done via introgression breeding (transferring genetic material from a species into the gene pool of another by backcrossing of a hybrid with one of the parent species), insertion of gene candidates and precision mutagenesis. CRISPR-Cas based systems can be used to knock out a gene, knock in a gene, and/or recombine genes at specific locations along the genome. It is something that is embarked upon not only by agricultural scientists, but environmental scientists looking for ways to mitigate the damaging effects of climate change.

The ability to use re-wilding as a strategy to combat climate change has been demonstrated in several plant varieties with the potential to become agrifood crops. For example, re-domestication of crop progenitors (wild relatives) in addition to the domestication of wild species has been demonstrated in *Solanum pimpinellifolium* (stress-tolerant wild tomato relative), *Physalis pruinosa* (groundcherry a distant relative of the tomato) and *Oryza alta* (wild tetraploid rice) by modifying domestication genes using CRISPR-Cas technologies (Cardi et al., 2023: 16). Other crops such as potato, kiwi, and pepper are the current focus of research into the ability to take genes from wild relatives to modify their modern, domesticated cousins. CRISPR has the potential to select appropriate genetic material to develop novel plant varieties combining good agronomic performance with adaptability to abiotic stresses and low input agricultural practices. The agrifoods listed in Table 4.1 are not typical staple agrifoods that populations depend on for survival, but research into re-wilding can yield important information about the potential to take traits from relatives of staple crops to make them more resilient to the effects of climate change, such as wheat.

Domesticated wheat has lost 70% of its genetic diversity compared with wild emmer (hulled wheat, a type of awned wheat) which had greater genetic diversity for abiotic and biotic stress tolerance (Haudry et al., 2007). Rice and soybean have lost 50% of their genetic diversity via domestication. Maize has lost 2–4% of its genes from the wild maize relative (*Zea mays ssp. Paviglumis*) through domestication processes (Razzaq et al., 2021: 6124). Some scientists argue that re-wilding could help combat the negative effects of climate change on agrifood production. It is a 'tool in the toolbox' that can be used when necessary.

Nevertheless, there are criticisms of the 're-wilding' argument that recognize some of the challenges with this approach to improving resiliency in important staple crops. Informant 8 (a private sector representative) expressed some skepticism in terms of the viability of what they called 'wide-crossed' crops. As they said,

> it still tends to be a last resort for plant breeders, and it would generally only be done when they've exhausted other options.

There is significant upstream work that would be needed to find desirable traits in wild relatives, and years of research that would be necessary to find appropriate genes to knock out, knock in, or recombine. Wild varieties may not yet have their genomes sequenced, which would be the first step to investigating what traits could be re-introduced into their domesticated relatives. The very first step to realizing the potential of orphan crops and re-wilding important agrifood crops would be to sequence the relevant genome, and experiment with various gene edits to develop promising plants that can adapt to future climate uncertainties.

4.4 Beyond CRISPR-Cas9

Several emergent platforms do not rely on CRISPR-Cas9 for gene editing. As Labant reports in 2022's Genetic Editing and Biotechnology News, a new system for gene editing known as ARCUS has been discovered by scientists. This consists of enzymes derived from I-*Cre*I, a shellfish genetic element that occurs in the algae *Chlamydomonas reinhardtii*. Other starting materials from nature include mobile genetic elements (MGEs), which have potential as "gene writing" tools that eschew double-strand breaks (Labant, 2022).

Other gene editing techniques do not rely on the Cas9 protein to manipulate the genome. MAD7 for example, is gaining popularity. MAD7 is a CRISPR enzyme that is similar to Cas9 and Cas12a.[4] Class II type V CRISPR-Cas12a is a new RNA guided endonuclease that has been recently harnessed as an alternative genome editing tool, which is emerging as a powerful molecular scissor to consider in the genome editing application landscape (Bijoya & Montoya, 2020). MAD7 is freely available from the patent holding company (Inscripta) and has been made available for use in research and commercial R&D so that it can be tested (Inscripta, 2021).

In 2021, the first successful use of MAD7 in a plant genome was discovered by a team of scientists at the Chinese Academy of Sciences in Beijing (Lin et al., 2021). An article in the *Journal of Genetics and Genomics* describes how the team use the MAD7 nuclease to genetically alter rice and wheat, demonstrating its potential for engineering crops. The results demonstrate the editing efficiency of MAD7 in rice and wheat. It is up to 65.6% successful in producing edits, comparable to that of the widely used LbCas12a CRISPR system. Despite being one of the most robust Cas12a nucleases, LbCas12a in general is less efficient than SpCas9 (another naturally present nuclease) for genome editing in human cells, animals, and plants. Cas12a nucleases, is thus called 'LbCas12a'. Additionally, the authors demonstrate that this approach can be used for multiplex gene editing when used with other CRISPR orthologues.[5] In the second publication in Progress in Molecular Biology and Translational Science, the authors discuss the MAD7 nuclease as an important resource that can overcome current limitations of CRISPR editing:

> A wide range of families and orthologues of CRISPR-associated proteins are being developed to fill the gaps in genome engineering by increasing their functionality, specificity, and […] and ease of access globally (Bayarsaikhan et al., 2021).

MAD7 was a topic of discussion for many of the research scientists we interviewed. Some seemed excited about the prospect of a CRISPR-based gene editing system that is not based on a royalty payment model, as MAD7 is free to use and develop gene edited products with commercial potential. As Informant 17 said,

> There are some limitations to MAD7 as far as it is based on its some of the biochemical properties…my lab's exploring that as well. Because of the unique licensing approach…academic labs, government labs would be able to develop traits that would allow small and medium companies to come up and run with them…So it's an important area of research but, CRISPR-Cas9 is the gold standard in terms of precision and accuracy.

Informant 5 was intrigued by alternatives to the Cas9 system, but pointed out that,

> there was path dependence, because the regulators, once they'd approved it 3 or 4 times, said 'we're good with that, we understand there's no risk with that promoter using this technology'. But every time you bring in a new promoter you got to go back to square one

---

[4] The Cas12a (formerly Cpf1) protein family involves 'Lb' (*Lachnospiraceae bacterium*) and 'As' (*Acidaminococcus* sp.). Cas12a is an engineered nuclease of Class II type V-A CRISPR-Cas (Cas12a/Cpf1).

[5] Orthologues are homologous genes that diverge after evolution gives rise to different species, an event known as speciation. The gene generally maintains a similar function to that of the ancestral gene that it evolved from. The point or event in evolutionary history that accounts for the DNA sequence variation within the gene determines whether the homologous genes are considered 'ortho' or 'para' (www.sciencing.com).

and prove the basic safety. You can invent a new one but it may not make any economic sense, because unless you can find one that you get enough volume out of that the people will invest in, taking to market will be just too expensive to reinvent. So it's cheaper to license from the other one.

Some platforms may be able to get better results, or different results, but may face regulatory hurdles if regulators are not familiar with the mechanisms of change.

The potential of CRISPR-Cas systems to be applied to transgenic crops is also being explored. Informant 12 (an academic researcher/social scientist) sees the future of gene editing as a useful application to transgenic crops to achieve multiple desired traits. They say,

> …a lot of gene editing is probably going to happen, like further transforming products that are already transgenic. There are a couple of traits that are extraordinarily valuable that you cannot get. They are from foreign DNA, for example, Bt. It is an extraordinarily valuable innovation and is from a bacteria. That's the whole point…maybe we'll find some equivalent of proteins or something. But that's not going away anytime soon, I don't think whatsoever. And there's herbicide tolerance. This is achieved through transgenic methods…My understanding is that there is not a knockout strategy for herbicide tolerance, and herbicide tolerance is also an extraordinarily economically valuable asset.

We briefly discuss three other approaches to altering the genomes of agrifood: base editing, prime editing, and RNAi technology. Base editing is an evolution of CRISPR-Cas systems. It does not require a Double Strand Break (DSB), which many proceeding gene editing technologies require. Base editing has emerged as an alternative tool to homology-directed repair (HDR) mediated replacement. Since there is low efficacy of HDR in plants, base editing can allow for precise nucleotide changes in the genome. Base editing facilitates precise editing of a plant genome by converting one single base to another in a programmable manner. For single-base substitution, base editing is emerging as an alternative and efficient powerful tool to HDR-mediated precise gene editing in plants. Base editing can be used to improve yields, nutritional qualities, pest resistance, and herbicide resistance in plants. Since base editing does not require DSBs or donor templates, they are a new way to think about ways to edit plant genomes in agriculture (see Li et al., 2023).

Like base editing, prime editing can make changes to a DNA strand without the DSBs. The prime editing system is derived from the CRISPR-Cas system. Prime editors, unlike other techniques, do not require DSBs or a donor DNA template to make gene edits. They require a programmable 'cutting enzyme' (nickase) such as the Cas9 protein, that is capable of cutting the DNA at a specified site. The nickase is fused to a polymerase enzyme which is an enzyme that synthesizes DNA or RNA and assembles DNA or RNA molecules. They can make almost any substitution, deletion, or insertion in the DNA of living cells. The feasibility of using Plant Prime Editing (PPE) in crops was demonstrated in 2019, but its efficacy was considered too low, especially compared with CRISPR-Cas systems. There is, however, significant current research into how to improve PPE's efficacy so it can be used effectively in plant breeding. Its efficacy has been demonstrated in herbicide resistance germplasm. There are current limitations to this approach, such as the inability for it to insert large sequences. There are also limitations of PAM sequences (Protospacer

Adjacent Motif is a 2–6 base pair DNA sequence following the sequence targeted by the Cas9 nuclease in the CRISPR bacterial adaptive immune system). Informant 16 (a research scientist) commented that

> …[there are] only a handful of labs in the world that actually tried prime editing. Prime editing might allow us to introduce very specific changes wherever we want them.

PPE has a lot of potential for precision editing and synthetic biology in agricultural plants, when research has revealed how to make it as efficient at editing for the desired traits as its counterparts (Tingting et al., 2023).

Finally, we turn to RNAi technology, which is an older technology relative to gene editing per se. The primary function of natural RNA interference is to regulate gene expression. RNA interference (RNAi) is a natural process found in most eukaryotic organisms (plants, animals, fungi) that was first identified in the late 1990s and was found to supress gene expression in a sequential order. Andrew Fire and Craig Mello won the 2006 Nobel prize for Physiology or Medicine for their ground-breaking work on how double strand breaks were responsible for gene silencing. Compared to CRISPR, RNAi technology reduces gene expression at the RNA level, what is known as 'knock-down', while CRISPR permanently silences the gene expression at the DNA level, what is called a 'knock-out'. Gene knock-down is important for research purposes because researchers can see the effect on a phenotype since reduced protein levels can be measured. Knock-downs are also reversable, so a researcher can return the protein expression to normal to observe the changes that the knock-down induced (Mezzetti et al., 2020).

One of the major downsides to using RNAi is off-target effects. CRISPR-Cas systems are much more precise and have less instances of off-target effects. Despite these challenges, researchers are experimenting with RNAi techniques to improve upon agronomic traits of plants such as targeting pest and pathogen genes within the plant's genome, as well as looking at surface applications. Decades of research using this technique has revealed target genes that can improve tolerance to biotic and abiotic stresses. RNAi technologies also have the capacity to down-regulate gene expressions without disrupting the expression of other genes. Though RNAi technology is much older than CRISPR-Cas systems it is another available tool in the toolbox for research scientists to help solve some of the most pressing challenges facing agriculture and food security.

4.5   Conclusions

The world of gene editing in agrifood is constantly changing. New discoveries are made everyday with the potential to revolutionize how we produce food and how the world feeds itself. In this chapter, we have discussed some of the newest breeding techniques being used in research labs around the world. It is important for regulatory systems to prepare for the wave of new breeding techniques on the horizon. Chapter 5 discusses regulatory change and the regulation of futures technologies.

References

Antle, J. M., Capalbo, S. M., Elliott, E. T., & Paustian, K. H. (2004). Adaptation, spatial heterogeneity, and the vulnerability of agricultural systems to climate change and CO_2 fertilization: An integrated assessment approach. *Climate Change, 64*, 289–315.

Atwell, B. J., Wang, H., & Scafaro, A. P. (2014). Could abiotic stress tolerance in wild relatives of rice be used to improve Oryza sativa? *Plant Science, 215*, 48–58.

Bayarsaikhan, D., Bayarsaikhan, G., & Lee, B. (2021). Recent advances in stem cells and gene editing: Drug discovery and therapeutics. *Progress in Molecular Biology and Translational Science, 181*, 231–269.

Bier, E. (2022). Gene drives gaining speed. *Nature Reviews Genetics, 23*(1), 5–22. https://doi.org/10.1038/s41576-021-00386-0

Bijoya, P., & Montoya, G. (2020). CRISPR-Cas12a: Functional overview and applications. *Biomedical Journal, 43*(1), 8–17.

Cardi, T., Murovec, J., Bakhsh, A., et al. (2023). CRISPR/Cas-mediated plant genome editing: Outstanding challenges a decade after implementation. *Trends in Plant Science, 28*(10), 1144–1165.

Coffey, D. (2020). What is a gene drive? *LiveScience*. Retrieved on January 2, 2024, from https://www.livescience.com/gene-drive.html

European Food Safety Authority (EFSA). (2023). *Proposal for a regulation of the European parliament and of the council on plants obtained by certain new genomic techniques and their food and feed, and amending Regulation (EU) 2017/625*. European Food Safety Authority (EFSA). Retrieved on December 19, 2023, from https://food.ec.europa.eu/system/files/2023-09/gmo_biotech_ngt_proposal_2023-411_en.pdf

FAO (Food and Agriculture Organisation). (2022). *Gene editing and agrifood systems*. FAO (Food and Agriculture Organisation). Retrieved January 3, 2024, from https://doi.org/10.4060/cc3579en

Fernie, A. R., & Yan, J. (2019). De novo domestication: An alternative route toward new crops for the future. *Molecular Plant, 12*, 615–631.

Gornall, J., Betts, R., Burke, E., Clark, R., Camp, J., Willett, K., & Wiltshire, A. (2010). Implications of climate change for agricultural productivity in the early twenty-first century. *Philosophical Transactions of the Royal Society B: Biological Sciences, 365*, 2973–2989.

Grain Research Development Corporation (Australia). (2023). *National varieties trials*. Retrieved November 13, 2023, from https://grdc.com.au/research/trials,-programs-and-initiatives/national-variety-trials

Haudry, A., Cenci, A., Ravel, C., et al. (2007). Grinding up wheat: A massive loss of nucleotide diversity since domestication. *Molecular Biology and Evolution, 24*, 1506–1517.

Hayes, K. R., Hosack, G. R., Dana, G. V., Foster, S. D., Ford, J. H., Thresher, R., Ickowicz, A., et al. (2018). Identifying and detecting potentially adverse ecological outcomes associated with the release of gene-drive modified organisms. *Journal of Responsible Innovation, 5*(Suppl 1), S139–S158. https://doi.org/10.1080/23299460.2017.1415585

Inscripta. (2021). *In new publications, scientists report use of MAD7™ nuclease for plant editing and more*. Inscripta Blog. Retrieved December 15, 2023, from https://www.inscripta.com/blog/in-new-publications-scientists-report-use-of-mad7-nuclease-for-plant-editing-and-more/

Labant, M. A. (2022, July 7). Non-CRISPR gene-editing platforms make the cut…or avoid making it. *Genetic Engineering and Biotechnology News*. Retrieved October 12, 2023, from https://www.genengnews.com/topics/genome-editing/non-crispr-gene-editing-platforms-make-the-cut-or-avoid-making-it/

Li, Y., Liang, J., Deng, B., Jiang, Y., Zhu, J., Chen, L., Li, M., & Li, J. (2023). Applications and prospects of CRISPR/Cas9-mediated base editing in plant breeding. *Current Issues in Molecular Biology, 45*(2), 918–935. https://doi.org/10.3390/cimb45020059

Lin, Q., et al. (2021). Genome editing in plants with MAD7 nuclease. *Journal of Genetics and Genomics, 48*(6), 444–451.

Lyzenga, W. J., et al. (2021). Advanced domestication: Harnessing the precision of gene editing in crop breeding. *Plant Biotechnology Journal, 19*, 660–670.

Mezzetti, B., Smagghe, G., Arpaia, S., Christiaens, O., et al. (2020). RNAi: What is its position in agriculture? *Journal of Pest Science, 93*, 1125–1130. https://doi.org/10.1007/s10340-020-01238-2

Montenegro, J. D., Golicz, A. A., Bayer, P. E., et al. (2017). The pangenome of hexaploid bread wheat. *The Plant Journal, 90*, 1007–1013.

Moriondo, M., Bindi, M., Kundzewicz, Z., Szwed, M., Chorynski, A., Matczak, P., Radziejewski, M., McEvoy, D., & Wreford, A. (2010). Impact and adaptation opportunities for European agriculture in response to climatic change and variability. *Mitigation Adaptation in Strategies for Global Change, 15*, 657–679.

NASEM (National Academies of Science, Engineering and Medicine). (2017). *Preparing for future products of biotechnology*. NASEM. https://www.ncbi.nlm.nih.gov/books/NBK442207/, https://doi.org/10.17226/24605

Naylor, R. L., et al. (2004). Biotechnology in the developing world: A case for increased investments in orphan crops. *Food Policy, 29*, 15–44.

Neve, P. (2018). Gene drive systems: Do they have a place in agricultural weed management? *Pest Management Science, 74*, 2671–2679.

Newman, S. J., & Furbank, R. T. (2021). Explainable machine learning models of major crop traits from satellite-monitored continent-wide field trial data. *Natural Plants, 7*(10), 1354–1363. https://doi.org/10.1038/s41477-021-01001-0z

Porter, J. R., Xie, L., Challinor, A. J., Cochrane, K., Howden, S. M., Iqbal, M. M., Lobell, D. B., & Travasso, M. I. (2014). Food security and food production systems. In C. B. Field, V. R. Barros, M. D. Mastrandrea, et al. (Eds.), *Climate change 2014: Impacts, adaptation, and vulnerability. Part A: global and sectoral aspects. Contribution of working group II to the fifth assessment*. Cambridge University Press.

Razzaq, A., et al. (2021). Rewilding crops for climate resilience: Economic analysis and *de novo* domestication strategies. *Journal of Experimental Botany, 72*(18), 6123–6139. https://doi.org/10.1093/jxb/erab276

Sinclair, T. R., Pinter, P. J., Kimball, B. A., et al. (2000). Leaf nitrogen concentration of wheat subjected to elevated CO_2 and either water or N deficits. *Agriculture, Ecosystems, and Environment, 79*, 53–60.

Tingting, L., Jinpeng, Z., Xi, Y., Kejian, W., Yuchun, R., & Chun, W. (2023). Development and application of prime editing in plants. *Rice Science, 20*(6), 509–522. https://doi.org/10.1016/j.rsci.2023.07.005

Tubiello, F. N., Soussana, J. F., & Howden, S. M. (2007). Crop and pasture response to climate change. *Proceedings of the National Academy of Sciences, 104*, 19686–19690.

Vikram, P., Swamy, B. P., Dixit, S., Singh, R., Singh, B. P., Miro, B., Kohli, A., Henry, A., Singh, N. K., & Kumar, A. (2015). Drought susceptibility of modern rice varieties: An effect of linkage of drought tolerance with undesirable traits. *Scientific Reports, 5*, 1–18.

Webber, B. L., Raghu, S., & Edwards, O. R. (2015). Is CRISPR-based gene drive a biocontrol silver bullet or global conservation threat? *PNAS, 112*(34), 10565–10567. https://doi.org/10.1073/pnas.1514258112

Wesseler, J., Kleter, G., Meulenbroek, M., & Purnhagen, K. P. (2022). EU regulation of genetically modified microorganisms in light of new policy developments: Possible implications for EU bioeconomy investments. *Applied Economic Perspectives and Policy, 45*(3), 839–859. https://doi.org/10.1002/aepp.13259

Wheeler, T., & von Braun, J. (2013). Climate change impacts on global food security. *Science, 341*, 508–513. https://doi.org/10.1126/science.1239402

Yaqoob, H., Tariq, A., Bhat, B. A., Bhat, K. A., Nehvi, I. B., Raza, A., Djalovic, I., Prasad, P. V. V., & Mir, R. A. (2023). Integrating genomics and genome editing for orphan crop improvement: A bridge between orphan crops and modern agriculture system. *GM Crops & Food, 14*(1), 1–20. https://doi.org/10.1080/21645698.2022.2146952

Zaid, A., Asgher, M., Wani, I. A., & Wani, S. H. (2020). Role of triacontanol in overcoming environmental stresses. In A. Roychoudhury & D. K. Tripathi (Eds.), *Protective chemical agents in the amelioration of plant abiotic stress: Biochemical and molecular perspectives* (pp. 491–509). Wiley.

Zhang, C., Jiang, S., Tian, Y., Dong, X., Xiao, J., Lu, Y., et al. (2023). Smart breeding driven by advances in sequencing technology. *Modern Agriculture, 1*(1), 43–56. https://doi.org/10.1002/moda.8

Chapter 5
Governing the Unknown: Regulating Future Technologies

> The big problem is the regulatory issues. It is the sticking point. Originally, I thought it might be the case with gene editing as well, but the regulatory issues seem to be sorting themselves out a lot more quickly than they ever did with transgenics.—Informant 20 (scientist/NGO representative)

5.1 Introduction

In this chapter we examine how governance structures and regulatory frameworks can respond to emergent new breeding techniques, exploring how emerging techniques may pose challenges for regulatory frameworks. We touch on the difficulties of governing the unknown, but also highlight the regulatory tools that may help policymakers better anticipate developments in a rapidly evolving regulatory space. We argue that using deliberative elements of governance will help stakeholders better identify opportunities and respond to future, unknown technologies. Finally, we take a closer look at the patent and licensing landscape for CRISPR technologies and how this space is in a state of flux. Can deliberative governance offer clues as to how to streamline and simplify the patent and licensing landscape?

We begin by discussing the principles of deliberative governance as it pertains to gene editing in the agrifood system. We then move on to what is currently changing in notable regulatory systems around the world with respect to emergent gene editing techniques and platforms. Finally, we take a closer look at the current patent and licensing landscape to provide a general overview of the legalities of innovation, and how this can help or hinder scientific advancements in gene editing applied to agrifood research and product development.

5.2 Regulating Future Technologies

One of the most influential documents published about the future of gene editing in the agrifood system is the National Academy of Sciences, Engineering and Medicine's report *Preparing for Future Products of Biotechnology* (NASEM,

© The Author(s) 2024
L. F. Clark, J. E. Hobbs, *International Regulation of Gene Editing Technologies in Crops*, SpringerBriefs in Environmental Science, https://doi.org/10.1007/978-3-031-63917-3_5

2017a).[1] Though the recommendations are directed toward US agencies responsible for regulating biotechnology, there are lessons to be drawn from NASEM meetings and subsequent publications that may be relevant to other countries (NASEM, 2024). Though much has changed since its publication, there are several actions outlined in the document that regulators today can implement to make their regulatory systems as effective as possible. The authors of the document cover the science behind gene editing but also the regulatory environment for gene edited agrifoods. In their Conclusions and Recommendations, the NASEM committee offers advice for how regulators could deal with future technologies as they move through the proof-of-concept stage towards commercialization. Those recommendations are worth examining here.

A key conclusion included in the final chapter of the NASEM document notes that:

> The risk-assessment endpoints for future biotechnology products are not new compared with those that have been identified for existing biotechnology products, but the pathways to those endpoints have the potential to be very different in terms of complexity. (NASEM, 2017a: 20).

Thus, regulatory frameworks may take different approaches to evaluating novel gene editing techniques or platforms, but the principles of safety assessment remain consistent with current ways of evaluating risks pertaining to human, animal, and environmental safety.

The committee observes that regulatory systems are likely not prepared for the number of emergent innovative technologies and applications on the horizon and may not have the necessary infrastructure, human resources or streamlined evaluation mechanisms in place. The committee recommends policy makers invest in public outreach and research and development within the agrifood sector. Outreach strategies that engage with stakeholders is an effective way of gathering information about what is in the pipeline, giving regulators some indication as to how to handle applications for approvals of novel agrifoods. Having an idea of what is in the pipeline gives regulators the ability to tweak the approval process to find a balance between stakeholder access to useful innovations, fostering economic development, while ensuring rigorous risk assessments to ensure the health and safety of humans, animals, and the environment.

Identifying and developing strategies for mitigating any potential harms while acquiring and maintaining social license and legitimacy are also issues policymakers need to consider. Not an easy task for any regulatory system, but the NASEM committee suggests investing in three primary areas that will enhance the regulatory ability to respond to emergent gene edited agrifoods. Broadly, the committee recommended a commitment of resources in key areas of expected growth of

---

[1] The US-based private, non-profit National Academies of Sciences, Engineering, and Medicine provides "independent, objective advice to inform policy with evidence, spark progress and innovation, and confront challenging issues for the benefit of society." For more information see https://www.nationalacademies.org/about

biotechnology, an increase investment in public and private research, and increase investments in regulatory science (NASEM, 2017a: Chapter 6 'Conclusions and Recommendations').

The committee's recommendations stress that increasing resources for policy, decision and social science research are necessary to improve the risk analysis processes that follow. Foresighting, scenarios and simulations have undoubtedly become easier with the advent of AI and big data to identify and predict potential bottlenecks and risks to health and safety before a product enters risk assessment. Whether regulatory agencies are using these tools is another question. They can be useful tools to help predict how a given regulatory framework may behave in the face of applications for approval of yet unknown gene editing applications. Foresighting is a systematic and purposeful process of future-oriented deliberation between actors with a view to identifying actions to be taken today for a better future tomorrow (Keenan, 2006, slide 4). These techniques can be used to adapt new foresight processes extending beyond the known risks to future issues and opportunities that face gene editing in agrifood. Foresighting can help identify future goals and criteria, as well as strategic opportunities.

The scenario method involves identification of drivers of change in a system, categorized by importance and uncertainties. These variables are used to form matrices of options for policy actions (Sharpe & van der Heijde, 2007). The NASEM committee also suggested that having a better understanding of the social, legal, and ethical implications associated with future gene editing techniques is crucial to having a responsive regulatory system. Consultations with stakeholders such as laypersons, as well as experts in the fields of public policy, innovation, agriculture, and gene editing may aid in this effort. It may also help improve the ways claims and counterclaims are handled to develop more inclusive approaches for determining "weight of evidence" in how data is interpreted (NASEM, 2017b: 37).

Improving science communication and knowledge exchange among stakeholders is foundational to improving preparedness for future advancements in biotechnology. On the practical side, bolstered field trials and pilot projects will provide real-world data through testing and information gathering. The last suggestion is to strengthen dialogue between regulatory agencies and trading partners to limit redundancies, and work towards policy harmonization to ease the regulatory burden and approve or reject emergent gene edited agrifoods as efficiently as possible to limit trade disruptions and costly misunderstandings, inaccurate or incomplete information.

This list of suggested changes and improvements to regulatory frameworks is no mean feat. Changing regulatory frameworks is costly. Allocating additional resources for future technologies takes resources away from other worthy issue areas for which governments are responsible. But there is a cost to maintaining the status quo and dealing with challenges on an ad hoc basis. It is important for regulators to decide how best to prepare for future uncertainties in the gene editing space that do not stall innovation, but also treat the inclusive, deliberative elements as equally important as the risk assessment of the actual organism or product. One notable point from the OECD (2018: 33) report on gene editing is that "what does not happen in one country, will likely happen in another." If governments are not prepared for innovation within their jurisdictions, they will have little say in how innovations develop abroad and land on their doorsteps with a possibly different set of standards and evaluations. Collectively, these are all approaches to help manage future uncertainties. They are also important components of deliberative governance, as defined by Hendriks (2009: 173), which in essence, is policy making that involves consultation with 'institutions, agencies, groups, activists and individual citizens coming together to deliberate a pressing social issue'.

5.2.1 Deliberative Governance

The suggestions from NASEM and the OECD largely reflect the principles for responsible governance laid out by Gordon et al. (2021) and include three areas: transparency and access; private and public management; and equity & inclusion. These are all elements of what we label deliberative governance, which puts effective, inclusive, and transformative communication at the heart of the governance of gene edited agrifoods. The 'transparency' aspects include publicly accessible information about the gene edited agrifoods that are being assessed for novelty, risk and health and safety; what the NASEM (2017a) report refers to as the 'data commons'. 'Private and public management' includes science-based regulation as well as voluntary best practices in conjunction with regulatory oversight. 'Equity and inclusion' cover risk avoidance and acquiring social license, as well as deliberative, inclusive social engagement and consultation.

Each of these components are equally important and deserve equal attention from regulators. Having a functional and effective regulatory system for gene edited agrifoods requires critical evaluation of what currently works in risk assessment platforms, and what needs to change in response to emergent technologies. Flexibility, and grounding evaluations on science-based information appear to be two key elements to effective systems that meet several overlapping objectives to approve or not approve products based on the scientific risks they may (or may not) pose to humans, animals, and the environment. Whether novel products stoke uncertainties regarding their economic implications once commercialized (who will benefit, who may not) or cause concerns regarding the scope of intellectual property rights in the food system, these issues are not considered relevant to risk-based assessments of novel organisms or products. Yet, questions regarding the socioeconomic implications of technological shifts are important and deserve attention within the deliberative governance approach.

5.2.2   The Precautionary Principle

A major issue is the future role of the Cartagena Protocol on Biosafety (CPB) in regulatory frameworks for new breeding techniques. Many countries have based their regulatory systems on the Precautionary Principle, which is a foundational part of the document. But there are limitations placed on regulatory systems adhering to the Protocol that do not necessarily reflect the realities of the risks inherent in using gene editing. Informant 5 (a research scientist) provided an insightful assessment of the ability for regulatory systems to reform when the CPB is embedded in their regulatory approach. Informant 5 observed,

> The issue is trying to come up with a way to move forward… countries find that they sign this international agreement on biosafety. And of course, they're also members of the Convention on Biological Diversity so that's meant that they have international obligations. And they built up this national biosafety framework and they tended to be kind of verbatim copies of Cartagena Protocol and in fact, the precautionary principle was in almost all of them. So, what that meant is that they based the national biosafety framework, their regulations, their laws, and their policies on this precautionary language, and part of the issue here is that they are stuck in a sense…what that means is that if you want to do something, you have to comply with that, and in some cases, particularly in Africa, that means a Parliamentary Act approved by Parliament and so trying to modify that means another 5 years, 10 years of discussions, until finally Parliament makes a decision.

The marriage of regulatory systems to the CPB may curb the ability for regulatory systems to assess emergent technologies accurately and effectively and be prepared for what lies ahead. Future breeding techniques will be subject to the risk assessments protocols designed for previous techniques and applications, that may not be relevant depending on the innovation. The lack of flexibility in strict precautionary approaches to evaluating new breeding techniques may stall innovation, restricting the ability for economies to be competitive and for producers and consumers to access useful innovations.

5.3   Change Afoot? The EU's Approach to Regulating Future Biotechnologies

In 2022, the European Parliamentary Research Service's Scientific Foresight Unit (STOA) published a document entitled, 'Genome-edited crops and the twenty-first century food system challenges.'[2] In it, the authors state that the emergent gene editing techniques (New Breeding Techniques (NBTs)) have the potential to meet the

---

[2] The Scientific Foresight Unit Network is active in international science and technology policy networks. It is a founding member of the European Parliamentary Technology Assessment (EPTA) network. The network works with European institutions and organisations, including the European Commission's Joint Research Centre (JRC) and DG Research and Innovation (DG RTD). For more information see https://www.europarl.europa.eu/stoa/en/about/stoa-network

objectives of the EU's Green Deal, specifically the Farm to Fork and biodiversity strategies. The authors were supportive of the reassessment of how gene editing agrifoods are regulated in the EU (EPRS-STOA, 2022). They did note, however, that the potential benefits from gene editing were often regarded by critics of biotechnology as hypotheticals and "achievable by means other than biotechnology" (EPRS-STOA, 2022: 22). This report was published before the European Commission (EC) proposal to reopen the dialogue regarding the regulation of NBTs similarly to GMOs. The proposal that was published in the summer of 2023 by the EC reinforced the points made in the early EPRS-STOA report and took the discussion a step further to re-open the debate regarding the EU's regulatory approach to gene editing.

With respect to risk to human and animal health and the environment, The European Food Safety Authority (EFSA) document outlining the proposed amendment to legislation pertaining to gene edited agrifoods concluded that there are no specific hazards linked to targeted mutagenesis or cisgenesis (EFSA, 2023). EFSA also concluded that in targeted mutagenesis, the potential for unintended effects, such as off-target effects, may be significantly reduced compared to transgenesis or conventional breeding. Therefore, due to how these novel techniques work, and compared to transgenesis, less data might be needed for the risk assessment of these plants and products made from them (EFSA, 2023: 2). The document states that the EU Directive 2001/18 which covers GMOs does not accurately reflect the risks posed by new breeding techniques using SDN1.

> The Commission… concluded that there are strong indications that the current European Union GMO legislation is not fit to regulate plants derived from using new breeding techniques obtained by targeted mutagenesis or cisgenesis, and products (including food and feed) derived from them and that that legislation needs to be adapted to scientific and technical progress in this area. (EFSA, 2023: 3).

Crucially, this document signals a need for reevaluation of how the current EU regulatory framework treats NBTs that have a different risk profile than GMOs (according to the committee), and that also offer profound potential to meet the EU's sustainability goals within the Farm to Fork Strategy. The proposal's 'general objectives' are very similar to the elements of deliberative governance as described by Gordon et al. (2021), including rigorous risk assessment to mitigate harms to humans, animals, and the environment, fostering innovation and economic competitiveness, and contributing to a sustainable agrifood system.

The proposal is now in the hands of member governments that must decide whether to proceed with proposed changes to the coverage of EU Directive 2001/18. There are several EU members that want a different set of regulations for NBTs, while there are also a number of members that wish to continue using the precautionary principle to evaluate NBTs in a similar fashion to GMOs.

Considering these recent developments and the speed with which this issue is progressing, it was important to include questions about the EU's proposal in our interviews. Informants had quite a lot to say about the EU situation and offered insights into how things may proceed now that it is up to member state governments to decide whether or not they are in favour of changes to how NBTs are regulated. US-based Informant 20 (a scientist working for an NGO) sees the proposal as an important first step, saying:

> Denmark has been rumored, for example, as changing policies and even France. I mean there's a lot of research going on genome editing in France and the President himself has come out and supported it. I think it's positive but it is going to take some time just because of the political structure.

Despite the optimism held by some informants that came with this announcement, Informant 11 (an academic researcher), who lives in the EU and researches issues related to agricultural biotechnology, had valuable insights into whether or not member states will approve this proposal. They seemed somewhat skeptical that this regulatory change would pass at the European Commission level, stating:

> Spain is pushing, the Netherlands are pushing, Sweden is pushing. But this is not enough… the problem with the NBTs is, that the Council would require a qualified majority by member states for any changes and, it's not likely that a qualified majority will be reached. Member states such as Germany have already abstained in the voting, so not voting in favor. Austria already has said that they are against any liberalization on the release of NBTs into the environment so they will vote no with France. And Italy can also be expected to abstain and just with 4 countries you cannot get a qualified majority anymore. A qualified majority requires that at least 15 member states vote in favour. And, that 55% of the population will be represented by those in favour. In France, Italy and Germany have already more than 55% of the population. And then under the EU rules these 3 big countries cannot block anything, you need at least a fourth country in the voting and that is Austria. They have already declared that they will vote against any liberalization. (Informant 11).

Outside of the EU there appears to be optimism, but inside the EU the perspective seems to be quite different. Informant 8 (a private sector representative based in the US) took issue with the arbitrariness of the 20 genetic changes that is the proposed threshold for whether a NBT would be subject to EU 2001/18. They said,

> So there are regulators who have said things like, if you make more than 20 nucleotide changes on the plant that has a regulatory consequence, and scientifically, that's nonsense. You can change one nucleotide and have a big effect, or you can change a thousand, and have not much at all. It's the result that matters.

Other informants had little faith that the EU would change how it regulates NBTs and stated that it would take a food security crisis for the hold-out member states to rethink their position. Informant 17 (a research scientist) commented,

> The only way perspectives on that will change in Europe for example is if there's true food scarcity or skyrocketing costs in Europe. That's the only way that they would change. And if it was recognizable that an edited crop would help alleviate that crisis, that's when it would make a difference. But many of the players that are driving their resistance to edited crops in Europe are very comfortable. Well, they have very comfortable lives and so they don't think it's an issue.

The direction the EU chooses to go has implications for other countries. Several Informants were critical of the influence of EU regulations on how countries in the developing world, namely in Africa, have developed their own regulations for gene edited agrifoods. As discussed in Chap. 3, many African countries are in the midst of developing regulatory frameworks, and some are working towards getting gene edited (and GMO) crops into the ground to reap the benefits of these revolutionary technologies. However, the negative influence of EU opposition to biotechnology in the domestic policy making space in other countries frustrated some of the Informants, most notably an Informant situated in an African country. Informant 19 observed,

> …you see that much of what is happening in Africa in sub-Saharan Africa, for example is influenced by Europe's stand on the issue. And if Europe is saying we are going to regulate edited products as GMOs that is going to have an effect on many countries in Africa and this is because of the complicated history of Europe and Africa. There is still that colonial idea, like if Europe has gone this way, Africa should also go this way and there's also the aspect of activities of very sophisticated activism for Africa to design its regulatory framework in a way that resembles those of Europe because it makes no sense because Africa trades with Europe more than it trades with the US…Europe's approach to this is going to have a ripple effect in much of Africa.

Several African countries are in the midst of developing harmonized guidelines that would cover any gene edited agrifood seeking approval in their jurisdictions. Kenya, Uganda, Tanzania, Rwanda, and South Sudan are in the latter stages of finalizing their regulatory frameworks for NBTs. Informant 19, who is contact with stakeholders in the above countries, said that harmonization of regulations may be the best way to move forward. They believe a move by the EU to differently regulate NBTs will have a positive effect on African countries' regulations, "whatever happens in Europe really has a major impact on what happens in sub-Saharan Africa." (Informant 19).

The world will have to wait and see what direction the EU goes in terms of how it assesses and regulates gene editing agrifoods. Various international organizations have offered suggestions as to how best to prepare for future new breeding techniques. As we discussed in Chap. 3 (Sect. 3.4), Canada has embraced the deliberative governance model in its handling of gene edited agricultural plants. Another significant issue to consider is the patent and licensing landscape that governs gene editing in the agrifood system.

5.4   Patents, Licenses and Freedom to Operate

The patent and license landscape for CRISPR and other gene editing technologies is complex, and continually changing. In the context of this book, we explain the basics of intellectual property law pertaining to patents, licensing and 'freedom to operate' as they relate to gene editing in the agrifood system. We also discuss some of the more substantive details of what entities are using gene editing technologies, like CRISPR, to conduct research and product development in the agrifood space. We discuss the race to file patented agrifood plant traits as regulatory burdens ease on the use of gene editing in the agrifood system and more gene edited foods become commercially available.

5.4.1 Intellectual Property Law and Patents: The Basics

Gene editing techniques are considered Intellectual Property (IP). Discoveries from using gene editing techniques are also considered patentable. Patents apply to techniques and the genetic traits that are outcomes of using these techniques. To protect IP, patents must be secured. Patents allow for owners to control others from making, using, selling, or importing the patented invention. Patents are territorial, and obtaining patent recognition is expensive and time consuming. They must be registered in each country or region the owner wants their property rights protected. Claim scopes may vary across countries or regions based on how the application is examined and the substantive laws in the country/region (Bagley & Candler, 2023). Before a patent is issued, the invention will be analyzed for novelty, invention step, proper description, and subject matter eligibility by the state or regional authority granting patent protection.

Countries have different legislation pertaining to what type of organisms can be patented. For example, in Canada, animals—even those with traits derived from gene editing—cannot be patented. The Supreme Court of Canada ruled on the 'Harvard Mouse' case in 2002.[3] Harvard argued that the *oncomouse*, which was genetically modified to have cancer-promoting genes for research purposes, required a patent in Canada. The Supreme Court ruled that higher life forms cannot be patentable, though in other jurisdictions (US, Japan) the Harvard Mouse was patented. There were worries at the time that this Supreme Court decision could chill biotech investment in Canada. This has not been the case.

Freedom to Use or Freedom to Operate (FTO) refer to the same thing. That is, the ability for an entity to use a patented technology. Freedom to Operate is defined as "the ability to proceed with research, development and commercialization of a product, while fully accounting for any potential risks of infringing activity, i.e., whether a product can be made, used, sold, offered for sale, or exported, with a minimal risk of infringing the unlicensed Intellectual Property Rights (IPR) or Tangible Property Rights (TPR) of others" (Kowalski et al., 2011: 12).

One cannot freely use a patented invention in a jurisdiction that does not have a patent on file without taking significant risk. Genome editing techniques, novel traits and relevant genes are patented in various jurisdictions around the world. Each country has its own set of criteria to judge the validity of a patent claim. Patent infringement liability may be claimed by the owners of the patent if such things were to occur. Other related patents may be filed in the country/region that could be used to prove infringement. If infringement is suspected, the burden of proof is on the accused infringer, as per the World Trade Organisation's Agreement on Trade Related Aspects of Intellectual Property (TRIPS), Section 5, Article 34. It states,

---

[3] Harvard College v. Canada (Commissioner of Patents), 2002 SCC 76 (CanLII), [2002] 4 SCR 4. See https://www.canlii.org/en/ca/scc/doc/2002/2002scc76/2002scc76.html

...if the subject matter of a patent is a process for obtaining a product, the judicial authorities shall have the authority to order the defendant to prove that the process to obtain an identical product is different from the patented process. (WTO, 1994).[4]

Though the WTO and the World Intellectual Property Organization (WIPO) administer treaties related to IP, they do not provide standards for the patentability of inventions.

Licenses are required to use patented genes/traits. The 2004 Supreme Court Canada (Schmeiser vs. Monsanto)[5] set the precedent that using plants with novel traits (derived from genetic modification or otherwise) require the user to pay a licensing fee regardless of whether the producer utilized the patented gene or not. Producers must pay licensing fees and pay royalties when required and gene edited plants use the same IP model as transgenics.

5.4.2   Who Owns CRISPR?

There have been several claims of inventing CRISPR, the most widely known of the gene editing techniques. Up until 2022, 'who owns CRISPR?' was unsettled. We would argue that this remains the case. After years of litigation, in February 2022 the United States Patent and Trademark Office (USPTO) named the Broad Institute as first to invent the use of CRISPR-Cas9 in eukaryotic cells.[6] The University of California Berkely (Dr. Jennifer Doudna), University of Vienna and Dr. Charpentier (CVC) dispute this claim, but today, the Broad Institute (of Harvard and MIT) legally holds the rights to the CRISPR-Cas9 patent in the US, though CVC holds over 55 patents related to CRISPR-Cas9 that are upheld. CVC holds the broadest foundational patents in the EU, Brazil, Canada, and Australia. The initial Broad patent for CRISPR-Cas9 in the EU has been revoked, with no appeals available. However, the European Patent Office (EPO) Opposition division in 2023 upheld three of Broad's patents for CRISPR-Cas12a (Sandys, 2023).

Obtaining a license for using CRISPR for commercialization is the path usually taken by developers. Licenses can include future royalties and milestone payments if a product is intended to be commercialized. However, obtaining a license, as Bagley and Candler (2023: 42) note, is not always straightforward. Patents are not published until 18 months after the earliest effective filing date. A potential user may not have knowledge of what entity or entities they must obtain a license from to use a specific CRISPR tool. Using patented techniques can also lead to discoveries. Researchers using licensed CRISPR tools may discover an invention for which

---

[4]The World Intellectual Property Organisation administers treaties on IP. The Patent Law Treaty came into force in 2005.
[5]Monsanto Canada Inc. v. Schmeiser, 2004 SCC 34 (CanLII), [2004] 1 SCR 902. See https://www.canlii.org/en/ca/scc/doc/2004/2004scc34/2004scc34.html
[6]Eukaryotic cells have a membrane bound nucleus. Organisms with these types of cells include animals, plants and fungi and many unicellular organisms.

they themselves would have to seek a patent. There are also licensing issues for tools that facilitate CRISPR (promoters, agrobacterium) that users will have to navigate before using the technique.

The licensing landscape continues to change, so keeping up with patents and licenses is required to avoid potential infringement (Bagley & Candler, 2023: 48). If US agrifood researchers wish to use CRISPR-Cas9 in their research and product development, they must get a license from the Broad Institute or Corteva (formally Dow DuPont Pioneer), as they both hold licenses for CRISPR-Cas9. If a researcher is in Canada, Brazil, Australia, or the EU, they must get a license from CVC. If a developer wishes to commercialize a product in a particular jurisdiction, they must obtain a license from the recognized license-holder, in addition to the license-holder where the R&D was conducted. There is no guarantee that obtaining a license from one entity will protect a researcher or developer from infringement in the future, as patents are continually filed regarding CRISPR techniques around the world.

In 2014, there were 90 CRISPR patent landscapes. As of August 2022, according to Swiss-based CENTREDOC (previously IPStudies) there were 15,000 patent families and over 400 licensing agreements. This number includes all CRISPR patents and licenses. Patents that apply to agriculture specifically are more difficult to pin down. The information available to us is from January 2021 where 1400 patents were filed for agrifood plant advances (CENTREDOC, 2023). In the case of CRISPR, there is currently a patchwork of patents with varying claims dependent on where the patent was filed. According to CENTREDOC, in 2023 there were 200-plus CRISPR patent families published every month. Most notably, the majority of CRISPR application patents in the agriculture as well as therapeutics areas originate in state-owned Chinese research institutions (CENTREDOC, 2023). According to Bagley and Candler's (2023: 38) analysis, the Chinese Academy of Agricultural Sciences (189), Chinese Academy of Sciences (112) and US-based Corteva (96) collectively filed 397 patents for agrifood plants using CRISPR in 2020 (as of January 2021). It is unclear currently whether either of the Chinese institutions have licenses for these patented discoveries, as they are either not published, or not accessible to those conducting Internet searches outside of China. It is also possible that some of the collaborators in these studies have licenses for jointly patented agrifood plants.

Corteva Agriscience (formally Dow DuPont Pioneer) and the Broad Institute control a significant proportion of the licenses for CRISPR-Cas9 for agrifood plants, as well as Cas12a and Cas12b. Each nuclease has a separate license, because the family of patents have different co-owners due to inventor collaborations. In 2017, they announced a partnership agreement that created a joint-licensing framework for CRISPR-Cas9 technology in agriculture. These include foundational CRISPR patents and products and techniques reliant on the Cas9 system. Corteva has the right to sub-license CRISPR-Cas9 patents for agrifood held by several entities, including the Broad Institute. Non-exclusive licenses for the IP for commercial agricultural research and product development are a core component of this agreement. Because of this, Corteva can offer licenses in a bundle, giving users rights to access several CRISPR-Cas9 technologies.

Academic and non-profit researchers can freely use the CRISPR-Cas9 technology without paying the licensing fee that commercial endeavors would be required to pay. The technology, however, is prohibited to be used for CRISPR gene drives, sterile seeds, or tobacco for human use (Broad Institute, 2017). Despite the dispute over the CRISPR-Cas9 patent, the creation of the licensing pool added some certainty over access to CRISPR-Cas9 IP for researchers and developers. But as Informant 7 (a research scientist) stated,

> ...CRISPR-Cas9 which is the most popular gene editing technique that belongs to the Broad Institute...and they are happy for people to use it for research purposes. But anything beyond that, it becomes a real challenge, and there's still a lot of uncertainty.

The Broad Institute and Corteva are not the only patent and license holders in the CRISPR field. Benson Hill Biosystems holds licenses to what it labels as "CRISPR 3.0". CRISPR-Cms1, which uses a different nuclease from Cas9, Cas12a or Cas12b, is owned by Benson Hill and it positions itself as offering a cost-effective alternative to CRISPR-Cas9. Cms1's simpler RNA structure allows for more efficient gene editing, according to the company's website. The company states that

> (the) uncertainty around the CRISPR intellectual property landscape presents a barrier to entry for innovators wishing to utilize genome editing solutions for those interested in accessing this powerful tool... Benson Hill aims to empower innovators with clear intellectual property rights and a licensing model that is transparent and simple. In the past year we've licensed our Cms1 nucleases in wide range of applications and fields, ranging from microbial applications to crops such as soybeans, wheat and rice. (Benson Hill Biosystems, 2019).

Bayer (Monsanto), BASF and Syngenta hold non-exclusive rights to the CRISPR-Cas12 licenses held by the Broad Institute for agricultural applications. Bayer is currently collaborating with the company Pairwise to use CRISPR-Cas9 for several crops, like wheat, canola, and soybean. Pairwise is part of the licensing agreement that allows the company to use and commercialize agrifood products developed using CRISPR-Cas9 techniques.

As mentioned in Chap. 2 (Sect. 2.2), other genomic techniques continue to be useful to agrifood researchers, such as TALEs. The 2Blades Foundation is a US-based non-profit that partnered with the scientists in Germany who first published on TAL Code in 2009. 2Blades grants licenses for the use of TALENs in agrifood plants for research and commercial applications. Informant 16 (a research scientist) was concerned about the IP issues related to CRISPR-Cas9. They observed,

> in terms of freedom to operate, it's very important for that generated technology to be widely available just because it won't be restricted to large companies that have resources but will be available to academia and will be available to government to small companies with good ideas, but that they don't have resources to go all the way through. To pay licenses fees for the technology. But that's when we talk about CRISPR technology there are other technologies like the TALENs.

According to the non-profit International Service for the Acquisition of Agri-biotech Applications (ISAAA) website, 2Blades claims that the licensing/use of TALENs technology generated 650 million USD in 2019 (ISAAA, 2023). TALENs has successfully been used to edit the genomes of soybeans, rice, potato, maize, and wheat. 2Blades has given a no-cost license to the International Rice Research Institute to help facilitate food security efforts (ISAAA, 2023).

The licensing landscape for CRISPR-Cas9 is continually evolving. Bagley and Candler (2023) discuss the current licensing of CRISPR technologies used in agriculture, providing a snapshot of which entity holds the licensing rights to CRISPR technologies, the types of licenses and the financial terms. As of 2023, the three main license holders are Corteva Agriscience, The Broad Institute and Benson Hill Biosystems. Corteva and The Broad Institute share licensing rights over CRISPRCas-9 for agricultural use. They have several types of licensing agreements including internal R&D, commercial seed and crop trait products and commercial licenses for other agricultural products. Both institutions charge licensing fees, including annual maintenance fees, commercial milestone payments and royalties. The fee structures are determined on a case-by-case basis. For example, academic and non-profit institutions seeking licenses for non-commercial use or R&D do not pay the same fees as for-profit entities. Similar license types and fee structures apply to The Broad Institute's licensing agreements for CRISPR-Cas9, CRISPR-Cas12a, and CRISPR-Cas12b technologies. Benson Hill Biosystems holds the license for the CRISPR-Cms1 technology. Benson Hill negotiates agreement individually based on economic potential and the financial circumstances of the institution applying to use the technology. Licensing may involve up-front fees, milestone payments and/or royalties (Bagley & Candler, 2023: 49).

Though the field is a bit less complicated today, hurdles do remain that create uncertainty for researchers and product developers seeking to use the power of CRISPR to develop new agrifood traits. Some have flagged the patenting of publicly funded research results as problematic, questioning why they are not freely available (Scheinerman & Sherkow, 2021). The fluid nature of the patent environment also concerns research scientists like Informant 7. They stated,

> at (their organization) they do allow us to use CRISPR-Cas9 for research, but I get the feeling that they are almost becoming a bit reluctant even for that. Just because they're unsure, our intellectual property officers are like, 'is it okay to use for research purposes?'…people have been using it for many years, but I think they just worry. And what happens if it goes to court again, and then someone else gets it, and then it's not okay and I think there's just a lot of uncertainty there. And commercialization, that's not easy. For us, we would need to go through some company that has a license to use it. So that is limiting as well, I'd say.

5.4.3   What Does the Future Patent/License Landscape Look Like?

Questions have been raised as to the patenting and licensing of techniques ('the process'), as opposed to organisms and products as is the case with transgenics (Gehrke, 2019). There are others who argue that patenting and licensing of CRISPR increases the concentration of power in the agrifood R&D sector. Several informants had quite a lot to say about 'freedom to operate' and the challenges faced by research scientists. Informant 3 (a former academic researcher) stated,

> I'm very concerned that with the IP environments, we're going to be locked again in a situation where only large companies that have the financial means will be able to commercialize product derived from these technologies. In Canada we do not have these large national and international companies like you see in Europe, like you see in US…But even on the public side, a lot of the innovation is generated by our universities from Canada, but we do not have the freedom to operate and, in fact, we cannot use CRISPR-Cas9 for the purpose to develop a new plant trait based in the environment in Canada.

CRISPR-Cas9 is one of over 100 variants of CRISPR enzymes. As such, there continues to be a race to secure IP exclusivity for newly discovered nucleases. Gene edited plants are much more difficult to trace, because there is no modified gene 'tag' that indicates whether or not an organism was developed using gene editing techniques like CRISPR. Detecting genome edited DNA sequences is far more difficult than tracing transgenic sequences. There are, however, organizations working on techniques to identify gene edited organisms. If a mutation is novel, it is possible to develop a test for detection. Analyzing seed samples and enforcing patent rights will become more important as more gene edited agrifoods become commercialized (the purpose of the Seeds Database proposed in the Canadian legislation covering gene edited agrifoods, see Chap. 3, Sect. 3.4). As for stacked traits, it is more difficult to protect IP. Holders of IP will have to rely on regulations to enforce their proprietary rights by using plant breeder's rights and trade secrets for commercial protection from infringement (CBAN, 2022). Paraphrasing one Informant, "it's difficult to track things that don't have markers (like transgenics). People are not sure how to monetize things that aren't necessarily enforceable."

5.5   Conclusion

This chapter has examined the principles of deliberative governance, along with current changes taking place in some regulatory systems. Several policy areas remain unsettled, and (at the time of writing) we have yet to see how and if the EU will modify its regulatory stance on gene edited agrifoods. The decisions made by the EU will have a ripple effect throughout the world, as countries in sub-Saharan Africa, for example, continue to work towards building robust regulatory systems that foster innovation, while giving smallholder farmers access to safe and valuable agrifood technologies.

In this chapter, we have also taken a closer look at the current patent and licensing landscape and examined some of the implications for researchers and regulatory frameworks as new techniques and applications continue to emerge around the world. Regulatory systems need to be prepared for whatever the future may hold in terms of gene editing. Governments also need to consider the shifting landscapes of patents and licenses that govern who gets to use certain technologies, what it costs, and the implications for commercializing a gene edited agrifood plant that could positively contribute to food security and climate change mitigation strategies.

References

Bagley, M., & Candler, A. G. (2023). CRISPR patent and licensing policy. In *Assessment of the regulatory and institutional frameworks for agricultural gene editing via CRISPR-based Technologies in Latin America and The Caribbean*. Genetic Engineering and Society Center. NC State University. IDB. https://doi.org/10.18235/0004904

Benson Hill Biosystems. (2019). *Enabling genome editing to be a truly empowering technology*. Retrieved October 24, 2023, from https://bensonhill.com/wp-content/uploads/2019/05/CRISPR-Nuclease-Portfolio-General.pdf

Broad Institute. (2017, October 18). *DuPont Pioneer and Broad Institute join forces to enable democratic CRISPR licensing in agriculture*. Press Release. Retrieved December 12, 2023, from https://www.broadinstitute.org/news/dupont-pioneer-and-broad-institute-join-forces-enable-democratic-crispr-licensing-agriculture

CBAN (Canadian Biotechnology Action Network). (2022). *Patents on gene editing in Canada*. Retrieved October 25, 2023, from https://cban.ca/wp-content/uploads/Patents-on-Genome-Editing-cban-March-2022.pdf

CENTREDOC. (2023). *CRISPR technologies CRISPR patent analytics*. Retrieved October 23, 2023, from https://web.archive.org/web/20230331061656/https://www.centredoc.swiss/en/publications/crispr/

EFSA (European Food Safety Authority). (2023). *Proposal for a regulation of the European parliament and of the council on plants obtained by certain new genomic techniques and their food and feed, and amending Regulation (EU) 2017/625*. EFSA (European Food Safety Authority). Retrieved on December 19, 2023, from https://food.ec.europa.eu/system/files/2023-09/gmo_biotech_ngt_proposal_2023-411_en.pdf

EPRS-STOA (European Parliamentary Research Service Scientific Foresight Unit). (2022). *Genome-edited crops and 21st century food system challenges*. EPRS-STOA (European Parliamentary Research Service Scientific Foresight Unit). Retrieved on December 19, 2023, from https://www.europarl.europa.eu/RegData/etudes/IDAN/2022/690194/EPRS_IDA(2022)690194_EN.pdf

Gehrke, L. M. (2019). Health law: Is gene editing patentable? *AMA Journal of Ethics, 21*, 1049–1055. https://doi.org/10.1001/amajethics.2019.1049

Gordon, D. R., Jaffe, G., Doane, M., Glaser, A., Gremillion, T. M., & Ho, M. D. (2021). Responsible governance of gene editing in agriculture and the environment. *Nature Biotechnology, 39*, 1055–1057. https://doi.org/10.1038/s41587-021-01023-1

Hendriks, C. (2009). Deliberative governance in the context of power. *Policy and Society, 28*(3), 173–184I.

ISAAA (International Service for the Acquisition of Agri-biotech Applications). (2023). *Pocket K No. 59: Plant breeding innovation: TALENs*. Retrieved December 15, 2023, from https://www.isaaa.org/resources/publications/pocketk/59/default.asp

Keenan, M. (2006). *An introduction to foresight. 2006 technology foresight training programme*. UNIDO Technology Foresight Training Seminar. Retrieved on March 18 2024, from: https://www.slideserve.com/fordon/an-introduction-to-foresight-powerpoint-ppt-presentation

Kowalski, S., Priess, M. R., Chiluwal, A., & Cavicchi, J. (2011). *Freedom to operate, product deconstruction, and patent mining: Principles and practices*. The Franklin Center for Intellectual Property, School of Law, University of New Hampshire. Retrieved October 13, 2023, from https://www.wipo.int/edocs/mdocs/mdocs/en/wipo_ip_wk_ge_11/wipo_ip_wk_ge_11_ref_3_kowalski.pdf

NASEM (National Academies of Science Engineering and Medicine). (2017a). *Preparing for future products of biotechnology.* https://www.ncbi.nlm.nih.gov/books/NBK442207/, https://doi.org/10.17226/24605

NASEM (National Academies of Science Engineering and Medicine). (2017b). *Human genome editing: Science, ethics, and governance.* E Glossary. National Academies Press. https://www.ncbi.nlm.nih.gov/books/NBK447273/

NASEM (National Academies of Science Engineering and Medicine). (2024). *About us.* Retrieved January 2, 2024, from https://www.nationalacademies.org/about

Organisation for Economic Co-operation and Development (OECD). (2018). *Conference on genome editing: Applications in agriculture.* Retrieved on December 2, 2023, from: https://www.oecd.org/environment/genome-editing-agriculture/

Sandys, A. (2023). *EPO upholds three Broad Institute patents for next generation CRISPR-Cas12a.* Juve Patent. Retrieved on October 24, 2023, from: https://www.juve-patent.com/cases/epo-upholds-three-broad-institute-patents-for-next-generation-crispr-cas12a/

Scheinerman, N., & Sherkow, J. S. (2021). Governance choices of genome editing patents. *Frontiers of Political Science, 3*, 745–898. https://doi.org/10.3389/fpos.2021.745898

Sharpe, B., & van der Heijde, K. (2007). *Scenarios for success: Turning insights into action.* Wiley.

World Trade Organisation (WTO). (1994, April 15). *TRIPS: Agreement on trade-related aspects of intellectual property rights.* Marrakesh Agreement Establishing the World Trade Organization, Annex 1C, 1869 U.N.T.S. 299, 33 I.L.M. 1197.

Chapter 6
What's Next for Gene Editing in Agrifood?

6.1 Introduction

We began in Chap. 1 by asking two questions. *Why are there so few gene edited agrifoods on the market despite the initial optimism that accompanied the Nobel Prize-winning discovery of CRISPR-Cas9 over a decade ago*? and *What governance challenges and opportunities will shape the future applications of gene editing in the agrifood system*? Throughout this book, we have explored various elements of gene editing in the agrifood system—the technology and its applications, the regulatory system, the intellectual property landscape, as well as innovations and emergent applications and platforms in the genome editing space. We have discussed the governance challenges facing regulators responsible for evaluating yet unknown, emergent innovations. It has been the goal of this book to provide the reader with ideas and insights into why gene editing in agriculture has not advanced as quickly as initially expected.

The heterogenous global regulatory landscape for gene edited plants, the various techniques that can be used to achieve gene edits, the complicated patent and licensing landscape, as well as the lack of consensus of what constitutes a 'living modified organism' and how gene edited agrifoods should be tracked and could be labelled are all factors contributing to the current precarious status of gene editing in the agrifood system. The future governance of gene editing in agrifood depends on how individual countries or trade blocs decide to proceed with regulatory definitions of gene edited agrifoods. It also depends on what emergent techniques are in the pipeline, and the extent to which they relate to current gene editing techniques like CRISPR or take another route to editing the genome with MAD7 or RNAi technology, for example.

© The Author(s) 2024
L. F. Clark, J. E. Hobbs, *International Regulation of Gene Editing Technologies in Crops*, SpringerBriefs in Environmental Science,
https://doi.org/10.1007/978-3-031-63917-3_6

6.2 Regulatory Systems in Flux

Many opportunities and uncertainties surround the future of gene editing. As we discussed in Chaps. 2 and 4, new nucleases with capabilities for gene editing are being discovered every year. What does this mean for the future of new breeding techniques in agrifood? There will be more options for researchers and developers to unlock beneficial traits in the genome of agrifood plants, and though not a focus of this book, also livestock. Gene editing applications for fungi and microbes are also emerging as important components of a sustainable, climate smart agrifood system. These are important advancements that have benefits, not only for producers, but also consumers and, more broadly, the planet. If we can harness gene editing to improve upon ways to make agriculture more sustainable and edit plants so that they are more adaptable to turbulent climactic changes, the abilities for smallholder farmers to be more productive on the land with fewer pesticides and/or fertilizers can provide a more regionally stable food source.

The major challenges lie in getting these technologies into the market, and to the people who can benefit from them the most, without being financially prohibitive. This is where harmonization of regulatory frameworks, even on a regional basis, as is the case in Latin America, South America and potentially in sub-Saharan Africa will be key. The complexity of the patent and licensing landscape further deters progress in the area of proliferation. At this point, there does not appear to be any consensus among patent and/license holders on how to proceed.

Europe could be a game changer if the European Union decides to regulate gene edited agrifoods differently than GMOs. The uncertainty surrounding the EU remains a major limitation to the applicability of CRISPR-Cas systems in Europe's agricultural sectors (Cardi et al., 2023: 18). But we have yet to see whether the recent EU proposal moves forward at the member-state level or is rejected. As discussed in Sect. 5.3, there are varying expert opinions on which direction the EU will go. Rejecting the new breeding techniques proposal may disadvantage the EU in terms of global competitiveness in agricultural products and the ability to address agricultural sustainability and climate change. The eventual direction the EU chooses will have implications for its trading partners, especially those in African countries like Uganda, Kenya and Tanzania that are in the process of finalizing their regulatory frameworks for gene edited agrifoods. Informant 19 (a private sector representative) said of smallholders in sub-Saharan Africa,

> …much of the activism against modern technologies is really out of touch with what farmers are facing. Farmers are not looking for GMOs. Farmers are not looking for gene edited technologies. Farmers are like, 'I'm grappling with this pest. Do you know any solution that can help?' I wish there was a model that would just allow farmers to have the opportunity to choose.

Canada, Argentina, and Japan appear to be regional leaders in the gene editing regulatory area, all using the case-by-case approach to the risk assessment of novel gene edited agrifoods while, in the Canadian case, embracing the deliberative governance approach. These models, and the principles they embody, may be what regulators in other jurisdictions look towards as they finalize their own regulatory frameworks for gene edited plants. The countries that have taken the 'product-based approach' may gain stronger agricultural trade ties with countries taking similar regulatory approaches. We have yet to see if taking this approach yields more commercialized gene edited agrifoods that have useful traits in these jurisdictions and beyond.

It is also unclear which direction China will go with its regulatory framework, and how that will or will not align with the patent and licensing situation in the country. Informant 12 (an academic researcher/social scientist) provided interesting insights into China's role in the race for dominance in the gene editing space, commenting that there is:

> …a lot of discussion in the international space that China has been intentionally obfuscating their intentions on moving forward as they get some mastery in this space…they're a market leader as they enter the space and have internal developer expertise. So as they enter that space, they're not just opening themselves up to a bunch of foreign entities entering the Chinese market. They're able to have a lot of expertise to service their own market and to be exporters…It's very, very paralyzing for the international regulatory community trying to anticipate what China is going to do.

We have yet to see what China's role will be in the global gene editing landscape, if and when licenses are issued for the CRISPR techniques patented by state-run institutions. It may turn out that the patented techniques are useful in certain contexts and not others. China may also become more influential in global standards setting for new breeding techniques and IP discussions compared to the EU if the 2023 proposal to revise regulations covering gene editing in the EU agrifood system are thwarted. The rest of the world is looking to these countries to see which direction they will take gene editing in agriculture.

6.3 Navigating the Patent Landscape

As discussed in Sect. 5.4, the patent landscape is complicated, and the dynamics of the sector are evolving rapidly. In terms of the patent and licensing landscape, harmonization that can simplify the environment for researchers is needed. Those experimenting with gene editing techniques need a degree of certainty about what IP legalities are relevant to their discoveries, which are not, and which might be in the future. The fact that successfully filed patents are publicized 18 months after they are approved is problematic for researchers and developers in this area, who may be violating a patent in the meantime without knowing it. There are calls to democratize the licensing landscape so that researchers in less developed countries can develop agrifood products that may benefit small-scale farmers, sustainability, and food security efforts within their own countries without paying hefty licensing fees and royalties.

There are also worries that the complicated IP issues surrounding CRISPR-Cas systems and other gene editing technologies will concentrate IP and research among only the largest multinational companies that can afford a team of legal experts in patent law, licenses, and royalty pay-outs. The Broad Institute's joint licensing agreement with Corteva is a step forward, but it is ad hoc and only relevant to the patents and licenses that these entities hold. Some argue that the revolutionary CRISPR technique should be liberalized and be freely available for use to advance science in agrifood research globally, without the financial limitations of patents, licensing, and royalty payments, especially for those seeking solutions to social goods such as addressing global food insecurity, climate change and sustainability in the developing world. Informant 18 (a research scientist with an NGO) sees this as a pressing matter, especially in the wake of the global supply chain disruptions during and after the COVID-19 global pandemic. They state:

> on global trade and the importance of developing crops for the developing world, I think when you have the supply chains being disrupted, people would like to look for supply chains that are more secure in terms of food. So, if your commodities were coming basically from one part of the world and the supply chains are completely disrupted, trade wars being one such example, you would look for policies which ensure that the production and productivity of commodities in those specific zones is not stopped. So gene editing will play a role.

Informant 8 (a private sector representative) was concerned about freedom to operate complications arising from the current state of IP in gene editing, and was also concerned by objections to gene editing in the agrifood system saying,

> I hope we get to a point where we can collectively recognize that the market is big enough for everyone, and choice is important, and we can make this work without either killing the organic sector or killing the rest of the sector by making technology so difficult to use that it's not possible.

One issue that remains unclear at this stage in the development of gene edited agrifoods is whether or not gene editing a registered variety of a plant will be legally recognized as the same variety. Is gene editing a particular variety of plant legally equivalent to the already registered variety, or is a gene edited variety legally different? This is an ongoing issue that is not an easy fix. It will take time, and employing aspects of deliberative governance, to get a clearer picture of how the legalities of commercializing gene edited agrifoods will evolve in the future.

6.4   Consumer Considerations

Though not a central focus of this book, it is worth commenting on consumer preferences and their role in the future of gene edited agrifoods. Consumer preferences appear to be much more significant in the development and efforts to commercialize gene edited agrifoods featuring consumer-oriented traits, compared with the development of GMOs which are argued to benefit producers (and IP rights holders) more so than consumers (Yang & Hobbs, 2020a). For some developers, there is less resistance to labelling gene edited products compared with the labelling issues that affect GMOs, because they have found that consumers are not as resistant to gene edited agrifoods as they are to genetically modified foods (using SDN3/transgenes) (Yang & Hobbs, 2020a, b). And in fact, some companies want labels on their food products indicating that they are products of gene editing. As Informant 9—a representative of a private company expressed to us in an interview (while displaying a prototype label of a gene edited food product ready for commercialization):

> …in terms of acceptance, acceptance will happen when we make products that consumers and people experience. I wanna show the back of the package…this is how we're talking about gene editing. So, you've got this better flavor by CRISPR. And then a QR code…there's been a real positive reaction.

In Kolondinsky and Lusk's (2018) study of labelling gene edited agrifoods, they claim that having simple labels on consumer products reduces opposition to the technology. The FAO's recent document on gene editing in the agrifood system states,

> …labelling rules should be framed in a harmonized global system based on transparent science-based consideration of risks, in which new traits in food would be included in a label if they represented a fundamental change in the composition of the food; production method would not be a mandatory labelling requirement (FAO, 2022: 24).

Labelling may act as a deterrent, and consumers may avoid products carrying a gene edited label. But as Informant 12 (an academic researcher) notes, gene editing in the food system may not be top of mind for most people, despite what researchers may think:

> I think the public has a very complicated relationship with food in lots of different attributes, and that potentially a plurality of consumers may not care either way, it's just not relevant to their lives…You can just choose not to eat this in theory, right? If it's labelled, you can choose not to eat this. If you're taking away people's ability to choose what they eat it gets more complicated… I think the public cares about this, but not nearly as much as regulators think they do.

Informant 9, who works in the private sector shared with us their experiences with consumer acceptance of gene edited agrifoods:

> …acceptance will happen when we make products that consumers and people experience…It's not driven by 'let me help you understand what this is and how we tap into natural'. And it's not GMO… this is how we're talking about gene editing… so you've got this better flavor by CRISPR…[what matters is] does it look good? What's in it? Color? Does it pop fresh, nutritious taste, right price? That's it.

The gene edited agrifoods that have already received approval, such as the GABA tomato in Japan and the Pairwise leafy greens in the US, have nutritional, agronomic and/or environmental benefits that traditional breeding methods would require decades to achieve. Informant 19 (a scientist with an NGO) noted that consumer preferences are important reasons why developing countries should embrace the possibilities of gene editing in the agrifood sector. They noted:

> …how crops are chosen is also because of the consumer preferences. So I would say, much of the research I know was happening when I was in Uganda was also related to when consumers complained about conventional beans taking too long to be cooked…We need beans

that can be cooked very fast without spending so much money on firewood. Consumer preferences should determine which kind of crops and which kind of traits.

Prioritizing consumer demands for agrifoods like beans with thinner skins so they cook faster could revolutionize the lives of millions and help reduce deforestation by requiring less cooking fuel. Informant 9 explained that:

> a more efficient utilization of the land helps to preserve diversity because you keep your footprint the same or less. If you're finding alternatives to having to use more broadly applied pesticides that have other effects because you've made the plant itself, you've tapped into that sort of natural, variability that it has to resist disease. Then maybe you help that way, too. The regulatory system should be set up to incentivize that.

6.5   Climate Change and Sustainability

The development of agrifood traits using gene editing technologies could include the needs of smallholder farmers to better meet UN sustainability goals which include addressing climate change. There is active research around the potential for gene editing to assist in achieving goals related to climate change, sustainability, and biodiversity. As Informant 7 pointed out:

> one of the things that we're working on in my group is to use gene editing to reduce methane emissions from cattle just by increasing oil production in forage crops. So little things like that have this downstream consequence that could be really very beneficial.

They continue,

> …trying to reduce methane emissions, increasing carbon capture potentially, there are all sorts of things to be done to try and make agriculture better for the environment and reduce the impact. I would say in terms of climate change and then, obviously a lot of people are doing work in terms of trying to improve adaptability of crops to climate change as well.

As to what stakeholders can do to move the needle forward in terms of acceptance of gene editing, Informant 1 (an academic researcher) stated,

> …I think we need to do more research to better understand from both the public and private sector what barriers are…preventing you from making an additional investments, and until we know those answers I think it's really tough to direct policymakers or anyone saying, 'Here's the policy changes that will increase our agriculture investments' because I don't think we fully comprehend what the public and private, and even the producer funding organizations, see as barriers to increase investment. And until we understand that I don't think we'll see much change in the dollars.

Using deliberative elements of governance, such as foresighting and scenario building, regulators are better able to address what relevant stakeholders perceive as barriers. Regulators can then see barriers from various viewpoints to determine where policy change can move the needle to maintain social license while fostering innovation, economic growth and making agriculture more sustainable through the use of new breeding techniques.

Informant 19 gave us a perspective from policymakers in sub-Saharan African countries that are developing regulatory frameworks for gene edited agrifoods. They urge scientists to take more consideration of the politics surrounding New Breeding Techniques in agriculture. They state,

> I wish [scientists] understood the web of the policymaking process and how regulatory frameworks are designed. Being a scientist, you tend to always see the world through the lens of 'everything should be based on science' and we know that it is not only about science. So that's one of the challenges. Also, scientists should take accountability. They tend to be so much absorbed in the science, forgetting that science does not thrive in just a vacuum. There is the social and political context in which these innovations will thrive in, and I wish scientists would think about the complexity around developing innovations that would thrive in the social political context in which it is designed.

6.6   Conclusions: Future Opportunities

Moving forward, what does the future look like for the governance of gene editing in the agrifood system? The goal of this book is to give readers a timely assessment of the technologies used to achieve gene edits in agrifood plants, and how regulatory systems are responding to these technologies. We have argued that the deliberative governance model is an approach that accommodates future yet unknown uncertainties, while maintaining rigorous health and safety protocols, risk assessments to protect human, animal, and environmental health, while acquiring social license. The Canadian case described in Chap. 3 (Sect. 3.4) highlights how elements of deliberative governance have been used to modify the regulatory environment for gene edited agrifoods. But there is much more to do. Better communication and attempts to understand the science are paramount for regulators, but it is also important for scientists to have some sense of how politics conditions what they do and what is possible for gene editing in the agrifood system.

In this book we have examined the complexity of fostering innovation in agriculture while governing the agrifood system. Various countries have chosen different regulatory paths, often a different balance between innovation, risk, health and safety, and where gene editing fits into climate change strategies, sustainable agriculture, and food security. As Informant 19 said when referencing discussions about GMOs in sub-Saharan Africa, part of gaining acceptance is that:

> …people do not oppose newer technologies or controversial technologies because they don't know it, but because of the assumptions that experts make. They think people are opposed because they don't know GMOs, so let's throw down all facts about GMOs and they will change their mind…It's very important [to] not make assumptions as experts…frame the engagement in a way that speaks to what ordinary people value.

We can learn lessons from previous experiences with biotechnologies like GMOs in how to better communicate benefits in a transparent, engaging, and compassionate manner. When trying to identify barriers and obstacles, it is important to ask, 'what do ordinary people value'?

In terms of steps forward, this book only scratches the surface of the economic, social, and policy issues that are part of the governance of gene edited agrifoods. More social science research is needed to identify the barriers and opportunities facing stakeholders flowing from the patent and licensing landscape. Ongoing, open dialogue needs to be facilitated by organizations directly involved in the science, policymaking and commercialization of gene edited agrifoods such as plant breeders, regulators, the private and public sector scientists, NGOs as well as community members. Future research could provide a better understanding of the environmental and nutritional potential of gene edited organisms like microbes and algae and how governance structures will address these applications. As the application of RNAi technology advances, policy makers need to figure out whether this technology will be regulated similarly to GMOs or gene edited agrifoods. Finally, there is a need for researchers to take a closer look at the role of Artificial Intelligence in synthetic biology and precision agriculture as they relate to gene editing platforms.

Many informants with whom we spoke were optimistic about the future of gene editing in the agrifood system. Informant 20, who works for an international NGO, said in regard to the future of gene editing,

> genome editing products have the potential to bring some pretty excellent traits to the food system and I think once you get these products out there and people are familiar with them and the concept of manipulating genes, whether you're putting a transgene in or you're just changing and tweaking the base, they will become less scary in general to the public and to some politicians that may not be well versed in the science.

Despite the challenges in navigating the complex and evolving regulatory landscape for gene-editing applications in agriculture, this sense of optimism drives scientific discovery. Widespread recognition of the need to address these regulatory barriers offers further cause for optimism.

References

Cardi, T., Murovec, J., Bakhsh, A., et al. (2023). CRISPR/Cas-mediated plant genome editing: Outstanding challenges a decade after implementation. *Trends in Plant Science, 28*(10), 1144–1165.

FAO (Food and Agriculture Organisation). (2022). *Gene editing and agrifood systems.* FAO (Food and Agriculture Organisation). Retrieved January 3, 2024, from https://doi.org/10.4060/cc3579en

Kolondinsky, J., & Lusk, J. L. (2018). Mandatory labels can improve attitudes toward genetically engineered food. *Science Advances, 4*, eaaq1413.

Yang, Y., & Hobbs, J. E. (2020a). The power of stories: Narratives and information framing effects in science communication. *American Journal of Agricultural Economics, 102*(4), 1271–1296. https://doi.org/10.1002/ajae.12078

Yang, Y., & Hobbs, J. E. (2020b). How do cultural worldviews shape food technology perceptions? Evidence from a discrete choice experiment. *Journal of Agricultural Economics, 71*(2), 465–492. https://doi.org/10.1111/1477-9552.12364

Appendix A: Methodology

A.1. Methodology for Data Collection

Interview questions were designed to collect data on the informants' understandings of recent advancements in gene-editing technologies used in agrifood supply chains; shifts in regulatory frameworks guiding gene editing technologies; and how policy changes may impact the use of gene editing in the global agrifood system. The list of potential informants was compiled based on an environmental scan for authorship of recent, comprehensive publications published by international organizations, academic publishers and regulatory agencies focused on governance and/or regulation of gene editing in agrifood. Email invitations were sent to potential informants which included regulators, representatives of international organizations, natural and social science researchers. Each informant was given a list of questions before the commencement of the interview (see Sect. A.3.). One set of questions was directed at regulators and representatives of international organizations, and the other set of questions for researchers in the field of gene editing (natural and social sciences). Questions about scientific techniques were included in the set of questions for scientists, while the other set included questions about regulations. There was significant overlap in questions as well.

The semi-structured interviews took place over a video conferencing platform between June and September 2023, each interview spanning about 60 min. Prior to recording, we received consent from all informants and, as per university ethics requirements, and gave assurance that all identifying factors will be suppressed in any publications including excerpts from interviews.[1] The interviews covered topics related to current gene editing technologies, advancements in the applications of gene editing technologies and the regulation of gene editing technologies applied to

---

[1] Informants' identities are protected, as per University of Saskatchewan Behavioural Research Ethics Board approval (# BEH 3922).

© The Editor(s) (if applicable) and The Author(s) 2024
L. F. Clark, J. E. Hobbs, *International Regulation of Gene Editing Technologies in Crops*, SpringerBriefs in Environmental Science, https://doi.org/10.1007/978-3-031-63917-3

agrifood. We identified relevant stakeholders by scanning relevant documents (public reports from international organizations and regulatory agencies, academic journal articles, news stories referencing gene editing and agriculture) for authorship (with publicly accessible contact information). Once interviews began, additional informants were obtained through snowball sampling. We reached out to those people who were named, by finding their contact information via Web searches.

These interviews took place after the Canadian Food Inspection Agency and the Plant Biosafety Office (PBO) decision to treat gene edited agrifoods that do not include transgenes as equivalent to the safety risks inherent in traditional breeding techniques in May 2022 (see Chap. 3, Sect. 3.4). The EU's decision regarding the regulatory status of gene edited foods was published (though widely publicized prior to the official release date) during the interview process in July 2023.

A.2. Key Informant Identifiers

I=Informant
I1: academic researcher (economist)
I2: research scientist/institutional representative
I3: former academic researcher (economist)
I4: academic researcher (economist)
I5: research scientist
I6: NGO representative
I7: research scientist
I8: private sector representative
I9: private sector representative
I10: research scientist
I11: academic researcher (economist)
I12: academic researcher (economist)
I13: academic researcher (economist)
I14: private sector representative
I15: regulatory representative
I16: research scientist
I17: research scientist
I18: research scientist/NGO representative
I19: private sector representative
I20: scientist/NGO representative

A.3. Master List of Semi-structured Interview Questions

There are four issue areas we are going to explore in this interview: regulation, gene editing technology, impacts and applications of gene-editing and communication issues.

1. Regulation/Organizational

   - Do you engage with policy makers? If so, what agencies? Can you describe the nature of engagement?
   - Do you think government regulations for gene-editing techniques are a 'done deal' or is there potential for some of the more restrictive policies covering the use of gene-editing technologies to be re-assessed?
   - What are some ways regulatory systems can respond to emergent genomic techniques?
   - Do you think enhancing regulatory science can facilitate better communication between scientists, the public and governmental agencies regarding novel technologies in agrifood?
   - Can you tell me your perspective on issues surrounding freedom to operate using gene-editing? Does the uncertainty surrounding freedom to operate create barriers?

2. Technology

   - Is gene-editing technology a superior genomic technique for plant breeding in agrifood research compared to others (ZFN, TALENs)? Is it going to replace other technologies, like GM?
   - How do gene-editing technologies enable other genomic technologies on the horizon? What role do you see for gene drives in agrifood research?
   - Where do you think gene-editing and other novel technologies are headed? How might regulatory systems respond to novel applications of CRISPR-Cas9 such as CRISPRoff?

3. Impacts and Applications

   - How are crops chosen for research purposes? How does funding (public, private or both) influence the research objectives?
   - Who stands to benefit from the use of gene-editing techniques in agrifood research and commercialized agrifood products? Who may not experience those benefits you describe?
   - Do you think that the war in Ukraine, current global energy, food and climate crises have caused regulators in some jurisdictions to re-think their policies regarding gene-editing in the agrifood system?
   - How can gene-editing be used to protect biodiversity and fight climate change?

4. <u>Communication issues</u>

- What do you wish research scientists (or regulators) knew about regulators and regulatory frameworks for novel technologies/gene-editing in agrifood?
- What is one thing you wish laypersons and the public in general knew about using gene-editing techniques in the agrifood system?
- Do you have anything else to add?
- Do you have anyone else in mind that we should talk to about gene editing?

Glossary

Allele   A variant form of a gene.

BES—Base editing system   Gene editing technology that can 'knock out' genes, repair errors in genetic code or fix mutations.

Cargo (trans)genes   Confers any trait that can be genetically linked to an engineered drive system and are required for a gene drive to work. They can be designed to confer chosen traits.

CPB   The Cartagena Protocol on Biosafety is a supplemental international agreement to the Convention on Biological Diversity. It outlines how to protect biological diversity from the risks of biotechnologies, like Genetically Modified Organisms (GMOs).

Cisgenesis   A type of genetic modification where donor and recipient are from the same species.

CRISPR—Clustered regularly-interspaced short palindromic repeats   A gene editing strategy system found naturally occurring in bacteria involved in immune defense. Bacteria use CRISPR-Cas9 proteins to cut the DNA of invading bacteria (Bacterial adaptive immune system). CRISPR-Cas9 revolutionized gene editing because it is cheaper, faster and more accessible to use than previous gene editing platforms. There are other CRISPR-Cas proteins used in gene editing such as cas12a.

DNA—Deoxyribonucleic acid   Complex of molecular structures. It is a double helix of nucleotides (molecules). Found in eukaryotic and prokaryotic cells. It carries genetic instructions for the development, growth, function and reproduction of all living organisms and many viruses.

DSB—Double strand break   Both strands of nucleotides in the double helix are severed. There are gene editing technologies that require double strand breaks to perform gene edits, such as TALENs or ZFN.

Enzyme   Proteins that are catalysts for chemical reactions.

© The Editor(s) (if applicable) and The Author(s) 2024   109
L. F. Clark, J. E. Hobbs, *International Regulation of Gene Editing Technologies in Crops*, SpringerBriefs in Environmental Science,
https://doi.org/10.1007/978-3-031-63917-3

Epigenome   How cells control gene activity without altering the DNA. Changes in DNA that do not alter the underlying sequence. Defects in these processes are often associated with disease.

Eukaryotic cells   Any cell or organism that has a nuclear membrane surrounding the nucleus. These organisms include protozoa, plants, fungi and animals.

GABA   Gamma-aminobutyric acid is a naturally occurring amino acid that works as a neurotransmitter. Studies have shown that GABA may help reduce anxiety, depression in humans.

Gene drives   Allow an edited gene on a chromosome to copy itself onto its partner chromosome during cell division.

GMO—Genetically modified organism   Any organism that has genetic material changed through genetic engineering that does not occur naturally by mating or natural recombination.

HDR—Homology-directed repair   When a donor DNA carries a desired change and has homology with the target site and is used to introduce this change at the cut site. Specific intentional insertions, changes or deletions can be introduced.

Homologous recombination   Genetic information that is shared among similar or identical nucleotides of single or double strands of nucleic acids.

Introgression breeding   Transferring genetic material from a species into the gene pool of another by backcrossing of a hybrid with one of the parent species.

LMO—Living modified organisms   Any organism that that has a novel combination of DNA as a direct result of genetic engineering. It is defined in the Cartagena Biosafety Protocol. Many countries include this definition in their domestic biosafety regulatory frameworks covering biotechnology.

MAD7   A CRISPR enzyme that is similar to Cas9 and Cas12a. A new RNA guided endonuclease that is an alternative genome editing tool.

Mutagenesis   Genetic information in an organism is changed by a mutation. It may occur in nature, or in a laboratory setting.

Nickase   An enzyme that produces a 'nick' on the targeting strand of DNA.

NHEJ—Non-homologous end-joining   Cut DNA is rejoined, but while doing this a few base pairs may be eaten away or added resulting in random small deletions (up to 20) or additions (a few base pairs) of nucleotides at the cut site.

Nuclease   Enzyme that cleaves bonds between nucleotides of nucleic acids.

Nucleotide   Molecules that are the building blocks of DNA and RNA.

ODM—Oligo directed mutagenesis   Non-transgenic (no foreign genes) base pair specific gene editing platform.

Oligonucleotide Mutagenesis   Technique allows for a mutation to be inserted in a gene at a specific site.

Orphan Crops   Orphan (minor) crops are crops typically not traded internationally but can play an important role in regional food security. For various reasons, many of these crops have received little attention from crop breeders or other research institutions wishing to improve their productivity.

Orthologous genes   Orthologues diverged after a speciation event, while paralogous genes diverge from one another within a species. Orthologous genes are homologous genes that diverged after evolution giving rise to different species, an event known as speciation. In this type of homologous gene, the ancestral gene and its function is maintained through a speciation event, though variations may arise within the gene after the point in which the species diverged.

Phenotyping   Observable traits or characteristics in an organism.

PPE—Prime editing   A 'search and replace' gene editing technique. The technique directly writes new genetic information into a targeted DNA site.

Prokaryotic cells   Cells that do not have a true nucleus. Examples are bacteria and archaea.

NBT—New breeding technique   Also known as new genetic engineering techniques, are a group of gene editing techniques that could advance the development of new traits in plants.

Polymerase enzyme   Synthesizes DNA or RNA and assembles DNA or RNA molecules.

PAM—Protospacer adjacent motif sequence   2-6 base pairs immediately following the DNA sequence targeted by CRISPR-Cas9 nuclease.

RNA   Ribose nucleic acid or ribonucleic acid—ribose nucleotides and nitrogenous bases that function in cellular protein synthesis. It turns DNA instructions into functional proteins. It is present in most of all living organisms and viruses. Replaces DNA as a carrier of genetic codes in some viruses. It is single stranded and has 4 bases. It is made up of nucleotides, similarly to DNA.

RNAi—RNA Interference   Natural mechanism for specific gene splicing in eukaryotic cells. RNA is used in sequence-specific suppression of gene expression by double stranded DNA.

SDN   Site-Directed Nuclease genome editing involves the use of different DNA, cutting enzymes (nucleases) that are directed to cut the DNA at a predetermined location by a range of different DNA binding systems. After the cut is made, the cell's own DNA repair mechanism recognizes the break and repairs the damage, using one of two pathways that are naturally present in cells.

SSN—Site-specific nucleases   Customized nucleases designed to cleave or bind to designated DNA sites that leads to double strand breaks.

TALEs—Transcription activator like effector   TALEN are fusions of TALEs. A TALE DNA-binding domain is fused to a DNA-cleavage domain, which create TALEN.

TALENs—Transcription activator-like effector nucleases   A gene editing technique that uses restriction enzymes that can be engineered to cut DNA.

Transgene   A gene that is transferred from one organism to another. It can happen naturally or in a laboratory setting (genetic engineering).

ZFNs—Zinc Finger Nucleases   Artificial restriction enzymes that facilitate targeted editing of a genome by creating double strand breaks (highly specific genomic scissors).

Index

© The Editor(s) (if applicable) and The Author(s) 2024
L. F. Clark, J. E. Hobbs, *International Regulation of Gene Editing Technologies
in Crops*, SpringerBriefs in Environmental Science,
https://doi.org/10.1007/978-3-031-63917-3